IS THERE LIFE ON MARS ?

The 20 Big Universe Questions

STUART CLARK

SERIES EDITOR
Simon Blackburn

Quercus

This paperback edition published in 2014 by
Quercus Editions Ltd
55 Baker Street
Seventh Floor, South Block
London
W1U 8EW

Hardback edition originally published by Quercus Editions Ltd 2010 as
THE BIG QUESTIONS: Universe

A CIP catalogue record for this book is available
from the British Library

ISBN 978 1 78206 947 8

10 9 8 7 6 5 4 3 2 1

Text designed and typeset by IDSUK (Data Connection) Ltd

Printed and bound in Great Britain by Clays Ltd. St Ives plc

Contents

Introduction

Questions in astronomy are invariably 'big'. Even the simplest query can lead down a winding path of investigation that results in a profound answer. This answer may well be accompanied by a mind-blowing revelation, and this is surely one of the subject's greatest attractions. The overwhelming size of the Universe, stretching across billions of light years of space and billions of years in time, and the unimaginable numbers involved in its description, provide a sense of awe in themselves.

When you stand at a truly dark site – in a desert or some other wilderness where the only light to be seen is coming from the stars above – the stars fill the sky in such profusion that even the most familiar of the constellations is difficult to pick out. Although there may seem to be countless stars, in fact the human eye can resolve about 3000 under pristine conditions. This is but the tiniest fraction of the total number of stars in the Universe. It has long been a cliché to say that the number of stars in the Universe is the same as the number of grains of sand on all the beaches in the world, but whilst there is indeed a staggering quantity of sand grains on Earth, this total is not nearly large enough. According to the latest estimates there are some 70 sextillion stars in the entire Universe; that is 70 thousand million million million, or a seven followed by 22 zeros. To pursue the comparison, this roughly equates to the number of grains of sand to be found on the beaches of 10,000 Earth-like planets.

This book attempts to answer questions that spring to people's minds about the wonders of the Universe. There are discussions of the exotic, half-glimpsed celestial objects such as quasars and pulsars, and the glorious close-up investigations of the nearby planets, such as Mars and Jupiter. There is a chapter devoted to those renowned celestial 'superstars' that retain their mystery no matter how many years pass by: the black holes. The question 'What is a black hole?' persists as the inevitable one as soon as anyone learns about my background in astronomy. Do not expect a complete answer here because even the experts don't have that yet. The study of black holes exerts a powerful allure; when a full understanding is attained, that knowledge will almost certainly propel astrophysicists to a new comprehension of the Universe as a whole.

Of the 19 other big questions discussed here, some have definitive answers after centuries of scientific effort; others are tantalisingly close to a solution; and some still remain utterly without resolution. These unsolved ones are perhaps the most captivating because they set the agenda for modern astronomy and cosmology. Regardless of our ability, or not, to fully answer it, each question tackles an important foundation stone in both our perception of the Universe and our efforts to appreciate our own place in its vastness; each question also delves a little into that special magic that we all feel a touch of when contemplating the Universe.

What is the Universe?
The human quest to know what's out there

The 'Universe' is what we call everything: every planet, every star, and every distant galaxy. It is vast beyond human comprehension, yet that has not stopped us trying to make sense of it all. Throughout history we have peered at it, measured it and studied it in the hope of one day understanding it. We have made some significant steps; but, lest we become complacent, the Universe always has a new surprise to throw at us, a new challenge to test our imagination

The urge to understand the Universe developed early in human history. Babylonian stone tablets dating back to 3000–3500 BC have been found that record the variable length of day throughout the year; the Chinese have written records of eclipses since about 2000 BC. Around the world are the remains of prehistoric structures that display striking astronomical alignments. The oldest of these is the 5200-year-old tomb at Newgrange, Ireland. At dawn on the winter solstice, the shortest day of the year, the rising Sun shines its beam through a passageway onto the floor of the inner chamber.

On Easter Island in the Pacific, seven of the hundreds of enigmatic statues face in the direction where the Sun sets on the equinox, when night and day have equal length. It has been suggested that the great Cambodian Temple Angkor Wat is aligned so that the Sun rises over the eastern gate on midsummer's day. The pyramids of Egypt are also thought to show alignment with the stars. Whilst none of these constitute an observatory in a scientific sense, they clearly demonstrate that the builders had an understanding of the motion of the Sun and stars.

The earliest astronomical observations were almost certainly used by ancient humans to set the calendar. The phases of the Moon defined the passing of a month, and the passage of the Sun through the sky defined both the length of a day and of a year. As the year progresses, the Sun rises and sets from different points on the horizon. Stonehenge, the well-preserved stone circle on Salisbury Plain in England, has a well-known solar alignment. On midsummer's day, the Sun rises over an offset monolith known as the Heel Stone. Initially Stonehenge was thought to be a temple to the Sun god, but some researchers have found other alignments between the stones and the Moon and suggested that it could have been a prehistoric observatory, perhaps used primarily for the prediction of eclipses.

The earliest cosmology

The Greek word *kosmos* means 'orderly arrangement', and from it we derive today's words cosmos and cosmology. Cosmology is the branch of astronomy that goes about answering that most fundamental of questions, 'What is the Universe?', by studying the way the Universe behaves, how it began, and how it will all eventually end.

Cosmology as a true science only really started in 1916, when Albert Einstein published his General Theory of Relativity (see *Was Einstein Right*?). Before this, astronomers lacked the mathematical framework in which to describe the behaviour of the whole Universe, and so pre-20th-century cosmology tended to be an uneasy mixture of speculation and religious sensibility. Ancient cosmology especially was usually inspired by religion and the assumption that Heaven was located somewhere above our heads, in space.

The Egyptians based their cosmology on the human reproductive cycle. They believed that the sky goddess Nut gave birth to the Sun god Ra every year and that the changing altitude of the Sun with the seasons was its gestation in Nut's star-studded body. The Sun was said to be reborn every winter solstice, and returned inside Nut through her mouth at the spring equinox. In this way, Ra continually recreated himself, making the Universe an eternal, self-sustaining entity.

Early civilizations told stories inspired by the patterns of stars in the night sky. They imagined the lines joining stars to form pictures of familiar or mythical characters. In Mesopotamia (modern-day Iraq) archaeologists have unearthed stone tablets and clay ledgers dating back to 1300 BC, which detail many such 'constellations' including the 12 signs of the zodiac. These zodiacal constellations were given special significance because they inhabited regions of the sky through which the Sun passed, and they were subsequently adopted by the Greeks – the Assyrian Hired Man and the Swallow became Aries and Pisces, for example, and the Goatfish and the Great Twins became Capricorn and Gemini. In Ancient Greece wandering minstrels would drift from village to village, recounting the star myths in exchange for food and lodgings. At the same time, philosophers would come up with their own fanciful tales to explain the nature of the Universe. One of the earliest was the philosopher Thales, in the sixth century BC. He put forward the idea that space was filled with water, in which the Earth floats, that earthquakes were caused by waves in this water, and the stars moved because they were caught in gentler currents.

The Greek astronomer Claudius Ptolemy, who lived in the first century AD, compiled a list of 48 constellations, but since not all the sky could be seen from Greece, the regions around the South Pole remained uncharted until intrepid astronomers ventured far from Europe during the 16th and 17th centuries in order to chart the southern stars. Other new constellations were also proposed to fill gaps in Ptolemy's classical sky map. Inevitably this led to arguments as astronomers disagreed. In England, Edmond Halley proposed a constellation called Robur Carolinum (Charles' Oak), after the tree in which Charles II had hidden from the Roundheads following the battle of Worcester. Whilst the King was delighted with the honour, some of Halley's fellow astronomers were not keen, and quietly discarded the constellation from their maps.

Much later, in 1922, things were finally put on a firm footing when the International Astronomical Union ratified 88 constellations with defined boundaries, mostly based on the Greek model. These were not the only aspects of Greek astronomy to pass into modern usage. There was one Ancient Greek in particular who was not prepared to tell stories about the stars, or speculate about them; he realized that the first step on the way to true understanding was to measure them. That man was Hipparchus and he defined a system of classification for the stars still in use today.

The brightness of stars

Even to a casual observer of the night sky, it is obvious that some stars are brighter than others. More than two millennia ago, Hipparchus meticulously compiled a catalogue of 850 stars, recording each star's position and ranking its brightness. He had no equipment to measure the brightness; he simply made his estimates by eye. The brightest stars he called 'first magnitude', the faintest he termed 'sixth magnitude', and the rest he ranked in the categories between. Amazingly, astronomers still use this seemingly crude magnitude system today, although modern measuring devices have extended Hipparchus's original six classifications. At the top end of the scale, the very brightest stars are now given negative numbers; at the other end, the stars that can only be seen with the aid of a telescope are assigned magnitudes with numbers often much higher than six. From the surface of the Earth, the best telescopes can detect stars of between 24th and 27th magnitude, but in orbit, above the distorting effects of the Earth's atmosphere, the Hubble Space Telescope can detect 30th magnitude stars. Each magnitude category is about two and half times brighter than the previous one, so 30th magnitude is about 3.5 billion (3500 million) times fainter than the naked eye can see.

But in measuring these perceived brightnesses, we are forgetting that the brightness of a luminous object is affected by its distance from the observer, as well by as the actual amount of light it gives out. Thus, a nearby dim star may well appear brighter than a highly luminous star far away.

This behaviour is governed by what is known as an 'inverse square law', which means that if the distance doubles, the intensity of the light drops to a quarter; treble the distance and the intensity drops to one ninth of its original value. To acknowledge this, magnitudes measured without any correction for distance are known as 'apparent' magnitudes. The 'absolute' magnitude is the brightness value that has been corrected for distance. The red star Betelgeuse – widely known because it can be pronounced 'beetle juice' – has an apparent magnitude of 0.58, but leaps up to -5.14 on the absolute scale. It is a truly bright star indeed but comparatively far away. On the other hand, because it is so close, the Sun has an enormous apparent magnitude of -26.7, the brightest object in the sky. However, when corrected for its proximity, its absolute magnitude is just 4.8. In other words, our Sun, for all its glory and importance in driving life on Earth, is nothing but a thoroughly average star.

Under the wandering stars

The Ancient Greek astronomers, with their dedication to detailed observations and recordings, have left us a rich legacy of knowledge about the stars. The nature of five particular stars, however, eluded them. They called them *planetes*, meaning 'wanderers', because of their movement across the sky from one night to the next, unlike all other stars which remained 'fixed'. From their Greek name you might correctly deduce that these *planetes* are in fact planets – our nearest five planets, Mercury, Venus, Mars, Jupiter and Saturn, which can be seen with the naked eye. The Greeks could have no concept that these were worlds in their own right, and imagined them to be gods, or at the very least emissaries of the gods, whose influence affected the fortunes of individuals on Earth.

Two of these wandering planets, Mercury and Venus, follow orbits between Earth and the Sun. So, viewed from Earth they stay close to the Sun and are only ever seen in the twilight sky. Mars, Jupiter and Saturn orbit the Sun further out than the Earth and can be clearly seen making their slow paths through the night sky. Many early astronomers became dedicated to tracking the motion of all the wandering planets so that their future positions could be predicted. This task was seen as important because when planets drew close to one another, their influences were thought to combine and magnify. Thus, conjunctions, as they were known, were significant events that needed to be predicted in order to cast horoscopes.

With the advent of the Christian era, the opinion was widely adopted that the motion of the Heavens would always remain mysterious because the sky was God's domain and mankind's puny intellect could never understand His omnipotent will. This perspective began to change in the first decades of the 17th century when Johannes Kepler distilled the movement of the planets into three mathematical laws of planetary motion (see *Why Do the Planets Stay in Orbit?*). This proved that the Universe could not only be measured, but understood.

At the same time, in Italy, Galileo Galilei was making discoveries that sparked our fascination with the wider Universe. In 1609, he raised his telescope and pointed it at the misty band of light that stretches across the night sky, known as the Milky Way. Through his basic telescope, tiny by today's standards, Galileo could see that the Milky Way was composed of a multitude of faint stars. This was a revelation to all, because it had been believed that the entire Universe contained only what could be seen with the naked eye. Now, however, Galileo had shown that there was far more that lay beyond unaided vision. This realization was the start of the centuries-long fascination, with each generation of astronomers developing larger and larger telescopes to see fainter and fainter objects, which continues to this very day. The largest optical telescopes in use now are fully 10 metres across, some 500 times larger than Galileo's original telescope.

Celestial neighbours

Today we know that the Sun is one star in a giant collection known as the Galaxy, which contains at least 100 billion stars arranged in a spiral pattern in a flat disc and orbiting a bulbous hub of even more stars. From our position in one of the spiral arms, we see the disc as a haze of myriad stars – the Milky Way. The centre of the Galaxy is towards the south, in the constellation of Sagittarius. If you could observe from a dark site in the southern hemisphere, you might be able to see the Milky Way widen into the vast star clouds of the Galaxy's central bulge.

The thickness of the Milky Way's stellar disc is estimated to be about 1000 light years, one light year being simply the distance that light travels in a year. According to laboratory measurements, light travels through a vacuum at approximately 300,000 kilometres every second (186 thousand miles per second), so in a year it travels about 9.5 trillion kilometres (5.9 trillion miles). This is the distance of one light year; using this unit means we can keep the mind-bogglingly large numbers a bit more manageable. In the galactic disc, the density of stars is about one star every four light years or so, but in the central heart of the Galaxy, some 25–30,000 light years away from the Sun, stars are densely packed and create an elongated bulge about 27,000 light years in diameter and 10,000 light years high.

In the Sun's immediate neighbourhood within the disc, there are 33 stars. By 'neighbourhood', astronomers mean stars closer than 12.5 light years. The majority of our neighbours are smaller, dimmer stars than our own Sun. Known as red dwarfs (see *How Old Is the Universe?*), these celestial minnows make up the largest population of stars in the Universe. Just two stars in our neighbourhood are similar in size to the Sun, and only one is greater: Procyon in the constellation Canis Minor is estimated to be twice the Sun's diameter and to contain one and a half times the Sun's mass.

Near the centre of the galactic bulge, the density of stars is 500 times greater than in our neighbourhood. If the Sun and its family of planets were suddenly placed in the centre of the Galaxy, there would be other stars, possibly with their own planetary systems, just ten times further away than Pluto. In reality, within our solar neighbourhood the nearest star is more than 5000 times further away than Pluto. At the very centre of the Galaxy, astronomers believe that the density of matter is so great that a black hole exists (see *What Are Black Holes?*).

And beyond our Galaxy?

As vast as it seems, the Galaxy is not the whole Universe. In the grand scheme of things, it is little more than a small island in an expansive ocean and there are innumerable other islands. Each one is a separate galaxy in its own right, containing anything from a few million to a trillion (a million million) stars. Galaxies come in three basic types: spirals, barred-spirals and ellipticals. The spiral galaxies are particularly beautiful, with their sweeping arms of bright young stars surrounding a central bulge of older stars. The barred-spiral galaxies, of which our own Galaxy is an example, are similar but with a pronounced elongation that connects the central bulge to the spiral arms. The ellipticals are totally different in appearance; they can be much larger than the spiral or barred-spiral galaxies and anything from cigar-shaped to perfectly spherical. There are some oddballs as well, known as the irregular galaxies, some of which may once have been spiral galaxies, and others that are truly disordered.

Among the largest and brightest galaxies, the spiral and barred-spirals account for about three quarters of the population. But there are also huge numbers of small elliptical and irregular galaxies spread across the Universe, referred to as 'dwarf galaxies'. When these are taken into account, the ratios reverse because tiny spirals are rare.

Although many galaxies seen are isolated, some of them are drawn together in clumps, attracted by one another's

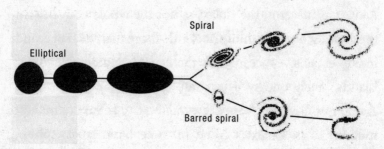

Types of galaxies

pull of gravity. At the smallest end of this behaviour, a collection of less than 50 galaxies is simply known as a group. Our Galaxy is part of the Local Group, which contains one other large galaxy – a spiral called the Andromeda Galaxy – and around 30 smaller galaxies. What we call 'clusters' of galaxies are essentially large groups containing more than 50 galaxies and in some cases more than a thousand. The nearest clusters to the Local Group are the Virgo Cluster containing about 1300 galaxies, the Coma Cluster with over 1000 and the Hercules Cluster that has around 100 members.

The various groups and clusters themselves gather into even greater collectives, appropriately named 'superclusters'. These vast collections are strung out through the Universe along giant 'sheets' or 'walls' known as 'filaments'. They appear to surround gigantic spaces containing scarcely any galaxies at all. If such voids can be thought of as celestial 'objects', then they are the largest objects in the Universe. A helpful way of visualizing the distribution of galaxies is to imagine the foam in a bubble bath – the galaxies are distributed on the thin soap film that surrounds the bubbles.

As for the number of galaxies that populate the Universe, with each passing year, the estimate goes up. Back in 1999, astronomers used Hubble Space Telescope observations to estimate that there were 125 billion galaxies. Not long afterwards, a new camera was installed on the Hubble and revealed many more, forcing the estimate to be doubled. Supercomputer estimates now suggest that there may be 500 billion (500,000 million) galaxies spread throughout the Universe.

Looking back in time

To investigate the origin of these galaxies and thus uncover the evolution of the Universe, cosmologists exploit the fact that light does not travel instantaneously across space. As fast as the speed of light is by our everyday standards - light could circle Earth's equator seven times in a second - it still takes many years to traverse the vast tracts of space between celestial objects. If a star is 100 light years away, its light takes 100 years to cross space to reach us and, as a consequence of this, we see it not as it is today but as it looked 100 years ago when the light began its journey. Cosmologists call this the 'look-back time'. It is like an archaeologist digging down through successive rock strata to uncover older and older fossils; the further an astronomer looks into space, the more ancient the celestial objects that appear in the eyepiece. With current telescope technology we can see celestial objects as they looked billions of years ago, and trace the way they developed into those we see around us today. Welcome to the world of the cosmologist.

How big is the Universe?
The cosmological distance ladder

Imagine that the distance between the Sun and Pluto is the length of a soccer pitch. The Sun would be a globe just 2 centimetres in diameter. The Earth would be 2.3 metres away from the Sun and just 0.2 millimetres across. At the other end of the pitch, Pluto would be nothing more than a speck of dust. Where would the nearest star be? - In the crowd? – In the car park? - In the next street? All wrong. The nearest star would be 645 kilometres (401 miles) away. And that is nothing on the cosmic scale.

Measuring the size of the Universe is unlike any other measuring job in human history. On Earth when you want to measure a distance, you can pace it out, or send a radar beam or laser ray to do the pacing for you. In the vast reaches of space this is usually impossible; the distances are simply too large. Bouncing laser beams or radar signals off celestial objects is only feasible for the Moon and the nearest planets. To measure other distances, astronomers have developed a filigree of different techniques for different distance ranges, called the 'cosmological distance ladder'. The variety of approaches is essential because no single distance determination method can serve across all scales of the Universe. Some celestial objects are too faint to be seen far away, others are too rare to be found nearby. Where the techniques do cover the same range, they serve to reinforce each other and improve the overall accuracy of the system.

Standard candles

Central to the cosmological distance ladder is the concept of the standard candle. This is a type of celestial object that releases the same amount of energy regardless of where in space it is found. Its distance is therefore the only thing that affects how bright it appears from Earth. One of the best types of standard candle is the so-called Cepheid variable star. The first example of such a star to be observed, Delta Cephei, caught a young astronomer's attention in 1784. John Goodricke of York charted the way it rose in brightness and faded again, deducing that the entire cycle took 128 hours and 45 minutes to complete. At this time astronomers knew of a handful of other variable stars, but each of those dropped in brightness suddenly and then restored themselves some time later. Delta Cephei was the only one to exhibit a gradual change.

By the first decade of the 20th century, many more examples of Cepheids had been discovered, some with shorter pulsation periods, some with longer. Henrietta Swan Leavitt, an assistant at the Harvard College Observatory, compiled a list of Cepheids in a nearby galaxy called the Small Magellanic Cloud. Rather than list them in random order, she wrote the 16 entries out according to how long they took to pulsate. Her curiosity was piqued by the fact that listed in this way the stars also appeared in order of average brightness: the longer the pulsation period, the brighter the star. By 1912, Leavitt had investigated a further nine Cepheids in the Small Magellanic Cloud and confirmed that each one's period of pulsation was tied to its average brightness. This immediately suggested that the Cepheids could be used as standard candles.

Final confirmation of this was supplied by British astrophysicist Arthur Stanley Eddington when he explained the behaviour of Cepheid variables. He proposed that the surface of the star trapped some of the outflowing radiation, causing the surface to swell up before releasing that radiation and shrinking again. What was more, he showed that the density of the star determined the period of pulsation. This means that all Cepheids pulsating with a period of say five and a half days will be identical to one another. Thus, any two can be compared and the difference in their observed average brightness can be used to calculate how much further away one is than the other. The comparison of Cepheid variable stars from one galaxy to another is one of the most heavily relied upon methods in distance determination.

A brighter standard candle is an exploding star, the super-nova type Ia. The explosion is triggered when the burnt-out core of a dead star siphons gas from a nearby, active star. The gas builds up on the surface of the dead star, increasing its mass. When this crosses a well-defined threshold called the Chandrasekhar limit (after Subrahmanyan Chandrasekhar, the Indian physicist who computed its value), the star can no longer support its own weight and collapses, setting off an enormous explosion. Because every supernova of this type is caused by the collapse of the central star when it reaches this mass limit, every one of these stellar cataclysms releases the same amount of energy into space. The brightness of a supernova type Ia is vastly greater than that of a Cepheid variable star; in fact it can outshine 100 billion (100,000 million) normal stars put together, rendering it visible across the entire Universe. The drawback is that these celestial detonations are impossible to predict because astronomers cannot see the doomed stars until they explode. Statistically, a supernova will occur about once a century in any given galaxy and this makes spotting one a task that requires constant vigilance.

Once a supernova type Ia has been seen, the distance of its host galaxy can be calculated relative to any other galaxy that has displayed a similar supernova. The same is true of galaxies seen to contain Cepheid variable stars. So, to join these two rungs of the cosmic distance ladder together, astronomers need to see a supernova in the same galaxy as they have seen a Cepheid. The more galaxies they can do this for, the better the accuracy of the join. Both methods give only relative distances, telling us for example that one celestial object is ten times farther away than another. What astronomers really want to know is the actual distance in kilometres or miles. To do this, they need to put a solid first rung on the cosmological distance ladder - one that supplies absolute distances. Luckily there is a method to do this and it is called 'parallax'.

The true distance to the stars

The orbital motion of the Earth around the Sun causes a seasonal shift in position of the nearest stars due to the effect of parallax. Astronomers began trying to observe stellar parallax during the 16th century, before the invention of the telescope. At the time, they were less concerned with measuring the distance to the stars than investigating the radical new proposal by Nicolaus Copernicus that the Earth was not immobile but in orbit around the Sun. They reasoned that if they could see the stars apparently shift position, they would be able to prove that the Earth moved.

Parallax can be easily demonstrated. Hold a finger in front of your face and close one eye. Notice the position of your finger in relation to some distant object, say a tree out of the window or a picture on the wall. Now change eye, keeping your finger stationary. Notice how the positions of your finger and the distant object have apparently changed. That is parallax, caused by the distance between your two eyes.

On the cosmic scale, stellar parallax occurs because the Earth is on the opposite side of its orbit every six months. As our viewpoint changes by almost 300 million kilometres (186 million miles) – the diameter of Earth's orbit – there will be subtle movements in the apparent positions of the nearest stars. The closer the star, the greater will be the movement. If the concept is simple, the execution is anything but. Despite many attempts, no parallax could be found in the 16th or 17th centuries. Not even Galileo's telescopes and their subsequent improvements allowed any hint of confirmation that the Earth moved. The reason was that the stars are so far away that their movements were too small to be seen

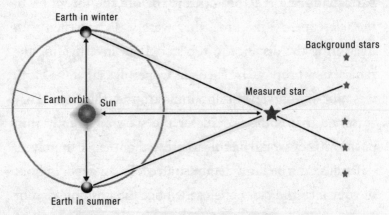

Parallax: the most accurate form of distance determination

by those early telescopes. Not until 1838 was the effect finally observed. Friedrich Wilhelm Bessel repeatedly measured the position of the star 61 Cygni and found that it displayed a minuscule parallax which allowed him to calculate (by trigonometry) that the star was around 93 trillion kilometres (58 trillion miles) away, or 9.8 light years.

Bessel's parallax measurement was swiftly followed by results for other stars from other astronomers. Despite this triumph, the work was painstaking and prone to error. By the first decades of the 20th century, astronomers had measured only about 100 stellar parallaxes. Today, satellite measurements have supplied parallaxes for over 100,000 stars, but all well within the Milky Way at distances of less than 1000 light years. Fortunately, some of these stars are Cepheid variables, so this has allowed the absolute brightness of Cepheids to be calibrated and the bottom rung of the distance ladder to be joined to the rest.

Redshift

In 1929, a discovery was announced that caused a revolution in cosmological thinking. American astronomer Edwin Hubble had found evidence that the Universe is expanding. As well as changing forever the way we think about the Universe, this provided another crucial rung in the cosmological distance ladder, allowing astronomers to measure distances across the largest imaginable swathes of space.

The seeds for Hubble's discovery were sown in the 1910s when astronomers recognized an exploding star in a fuzzy celestial cloud. These fuzzy clouds were known as 'nebulae' and at the time most astronomers thought that they were pockets of gas within our own Galaxy. Upon sight of the supernova explosion, some astronomers began to voice a very different opinion, that these nebulae were highly distant collections of stars - a view that became known as the 'Island Universe' hypothesis. Others clung to the status quo, maintaining that the nebulae were nearby. The debate was settled in 1924, when Hubble succeeded in finding a Cepheid variable star in the largest of the clouds, known then as the 'Andromeda Nebula', while using the 100-inch telescope on Mount Wilson in Southern California. He calculated the distance to Andromeda as 900,000 light years. Although we now know that the Andromeda Galaxy, as it was renamed, is 2.2 million light years away, Hubble's much lower estimate was at that time three times larger than the accepted size of the *entire* Universe. It proved not only that the Universe was vastly larger than previously believed, but that individual galaxies were dotted throughout space and that each one of these galaxies was a huge conglomeration of stars.

Spurred on by this, Hubble began a systematic study of the galaxies. He placed them into three classes: spiral, barred-spiral and elliptical galaxies (see *What is the Universe?*). He estimated the distance to as many of them as he could, and he studied the light from them. The key to his final discovery was noticing that the light from each galaxy displayed a 'redshift'. In 1842, Christian Doppler had proposed that if the distance between a source of light and an observer were shrinking, the light would be 'squashed'. This would shorten the wavelengths and so turn the light bluer, because blue light has a shorter wavelength than red light. In the opposite situation, when a source and observer are moving apart, the wavelengths are lengthened and the light becomes redder. This stretching is what we call redshift, and Hubble became an expert at measuring it in the light from galaxies.

Hubble plotted the distances of 46 galaxies against their redshifts, and saw something amazing: the further away the galaxy, the greater its redshift. According to the Doppler effect, this meant that the further away a galaxy was from the Milky Way, the faster it was receding. It was as if everything was exploding away from everything else in space. At a time when the prevailing view was that the Universe was static, this seemed incredible.

Expanding space

In 1916, several years before Hubble undertook this work, Albert Einstein presented his General Theory of Relativity to the world. General relativity provided the mathematical framework within which to properly understand the meaning of Hubble's subsequent observations. It too proposed that the galaxies were separating from one another, but rather than individual galaxies speeding through space, its premise was that space itself was expanding, carrying the galaxies along with it. A useful way to visualize this is by thinking of a raisin cake mixture. Before rising in the heat of the oven, it is a small dense lump with closely packed raisins. After it has risen, the raisins have not moved through the mixture but they have been driven apart by the expansion of the cake. So it is with the galaxies; space itself is endowed with the ability to expand and, as it does so, it drives the galaxies apart. The more space between a pair of galaxies, the faster they are driven apart and the greater the redshift. When astronomers talk about a redshift of 1, they mean that the light has been doubled in wavelength. To do this, the Universe must have doubled in size while the light has been travelling along. A redshift of 3 means the light's wavelength has quadrupled, hence the Universe has doubled in size twice, meaning it is now four times as big as when the light started its journey.

Getting back to our aim of measuring cosmological distances – if we know a galaxy's redshift, we can convert it to a distance once we know how fast the Universe is expanding. The expansion rate of the Universe is known as the 'Hubble constant', but measuring it confounded astronomers for most of the 20th century. Hubble himself got it wrong by a factor of almost seven. Although the 100-inch telescope he was using was then the largest in the world, it was incapable of resolving Cepheids throughout his sample of galaxies. So, although Hubble showed us how to measure the size of the Universe, the technological restrictions of his day meant that he could not complete the job – not even when he started using the 200-inch Hale telescope on Mount Palomar, California.

In the last two decades, astronomers have had another Hubble – the Hubble Space Telescope – which has allowed them to finish the task, thanks to its vantage point above Earth's atmosphere. Although Hubble's mirror is only about half the size of the Hale Telescope, it has identified Cepheids out to around 60 million light years and supernovae type Ia across billions of light years. It has enabled astronomers to calculate a definitive figure for the Hubble constant, which tells us that for every million light years that separates two galaxies, they are driven further apart at a speed of 22 kilometres per second (14 miles per second).

How big is space?

Using the now well-constructed cosmological distance ladder, astronomers today are confident that the Universe stretches on for billions of light years in all directions. The most distant celestial object yet observed has a redshift of just over 8. This means that the Universe has doubled in size eight times during the time it has taken for the light to reach us. It converts into a distance of around 13 billion light years, and this is our current estimate for the minimum possible size of the Universe. It has taken the light from that celestial object 13 billion years to cross the Universe and in all that time the Universe has continued to expand.

To calculate the *full* extent of the Universe, the only recourse is to create computer models of the expanding Universe based upon the laws of general relativity. These suggest that during those 13 billion years, the Universe has swollen to at least 95 billion light years in diameter. As if that were not brain-bending enough, there is one final sting in this tale. The Universe may extend way beyond 95 billion light years, but there is no way for us to see beyond even 13 billion years because there has not been enough time since the origin of the Universe for the light from such distant regions to reach us. If we wait another billion years, then the light from the next billion light years of space will arrive and we will be able to study it. Like a television soap opera, the study of the Universe seems destined to be a never-ending series of revelations.

How old is the Universe?
Cosmology's age crisis

No one needs an astronomer to tell them that it is dark at night. Apart from the stars that dot the sky, the darkness is the night sky's defining quality. Add an astronomer to the picture, however, and he or she will tell you that the darkness is actually one of the most profound observations you can make. It leads us to the idea that the Universe cannot have existed forever.

In astronomical circles the question 'Why is it dark at night?' is known as 'Olbers' Paradox', after German astronomer Heinrich Wilhelm Matthäus Olbers who popularized the discussion in 1823. Back then, the dark night sky was considered a paradox because the prevailing view was that the Universe had always existed and was infinite in size; stars were scattered throughout this unimaginable expanse and so in whatever direction we look, our line of sight should eventually intercept a star. Hence, the night sky should be starlit and as bright as day, not dark. Olbers was not the first to point out the problem. Before him, Thomas Digges in 1576, Johannes Kepler in 1610 and Edmond Halley in 1721 all puzzled over the darkness at night. It just did not fit their pictures of an infinitely old and vast Universe populated by a uniform collection of stars.

Various solutions were contemplated. It was a familiar fact, for example, that if you double the distance of a light source, you quarter its intensity (the inverse square law, see *What is the Universe?*). This means that distant objects will naturally fade from view. But this could not solve the paradox because the further you look the more stars seem to crowd together, compensating for the drop in individual brightness. So, astronomers were left with the paradox.

The solution is to let go of the assumptions. If the Universe is not infinitely old, then the light from more distant stars would not have had time to reach the Earth. If the Universe is not static but expanding, this would have a weakening effect on distant light because of the redshift (see *How Big is the Universe?*). If the stars are not spread uniformly through space but are corralled into galaxies with voids between, this would affect the 'line of sight' hypothesis. And we now know that there will never be enough stars in existence to fill the Universe with starlight because stars do not form fast enough or live long enough to release enough energy into space. Let us concentrate on the fundamental idea that the Universe is not infinitely old. This certainly helps to solve Olbers' Paradox but it leaves us with a greater mystery: if the Universe is not infinitely old, how old is it?

The age of the Earth

In trying to work out the age of the Universe, an obvious starting point is that the Universe cannot be younger than the objects it contains. There have been various attempts to estimate the Earth's age. The earliest were biblical in origin; theologians assumed that Man had always populated the planet and so counted the number of generations recorded in the Bible and used this to arrive at an estimate of the Earth's age. By the 19th century such attempts had given way to more scientifically based methods, for example deducing the time it would take for the planet to cool down, supposing that it had solidified from a molten mass. The discovery of radioactivity in the late 19th century, by Antoine Henri Becquerel and independently by Marie Curie and her husband Pierre Curie, signalled the death of this approach because radioactivity in rocks provides a constant source of heat and so ruins the calculations of the cooling rate. However, far from demolishing our way to age the Earth, it was soon realized that radioactivity actually gave us the very best route of all.

Radioactive elements decay from one element into another, giving out energy in the form of particles or rays. Different versions of the same element are called isotopes, and each radioactive isotope has a unique decay time. This is termed the 'half-life' of the isotope because it measures the time taken for half the quantity to decay. After some ten half-lives there is hardly any of the original substance left. Some isotopes have relatively short half-lives (on a geological scale), such as carbon-14 at 6000 years. One isotope of uranium, uranium-235, decays with a half-life of 704 million years, into lead; uranium-238 has a half-life of 4.47 billion years, and decays into thorium. By analysing a rock for its naturally occurring proportions of uranium, lead and thorium, geologists can estimate its age. If a wide enough range of samples is analysed, this technique can give the definitive age of the Earth.

After a century of effort, geologists have now radioactively dated myriad Earth rocks, Moon rocks and meteorites. This has given an age, not just for the Earth but also for the whole Solar System, of 4.6 billion years, and represents a preliminary baseline for the age of the Universe. As it turns out, it is somewhat on the low side.

The age of the stars

Looking further afield, isolated groups of stars known as 'globular clusters' provide excellent candidates for estimating ages. Each large galaxy has a retinue of globular clusters; there are more than 150 known to be orbiting the centre of the Milky Way Galaxy. The largest elliptical galaxies can possess more than 500. By studying the distribution of globular clusters, Harvard astronomer Harlow Shapley deduced in the early 20th century that the Sun was located far from the centre of our Galaxy. Had we been in the centre of the Galaxy, the globular clusters would have been distributed evenly around us. As it is, most of them gather tightly in the southern sky, revealing that we are seeing them from somewhere off-centre.

Each globular cluster is a spherical collection of a few hundred thousand stars. The types of star it contains can indicate its age. Most stars in the Universe are called 'main sequence' stars and this means that they are in the stable 'middle-aged' phase of their existence. A star's individual lifespan, and its characteristics, depend upon the amount of mass it contains. High-mass stars generate the most energy, have the hottest surfaces at more than 40,000 degrees Celsius, and give off blue-white light. They also have the shortest lifetimes - although a massive star has more fuel to burn, the extra weight pressing down on its core drives its nuclear reactions more quickly, so it runs out of fuel much faster than a smaller star. The highest-mass stars that astronomers have seen are about 100-200 times the mass of the Sun and are calculated to burn their fuel so furiously that they live for only a few million years. Less massive stars, conversely, generate energy more slowly and so live longer lives. They also have cooler surfaces and exhibit a different colour. The Sun is a yellow star with a surface temperature of about 6000 degrees Celsius and it is estimated that it will live for approximately 10 billion years. The least massive main-sequence stars are the 'red dwarfs'. With surface temperatures of less than 3000 degrees Celsius, they are so miserly with their nuclear fuel that calculations show the smallest of them may last for 100 billion years.

In an isolated stellar population, if no new stars are born the population becomes skewed towards low-mass stars because the higher-mass stars have lived and died off. Globular clusters are just such isolated systems. They formed their stars in one fell swoop at some point extremely far

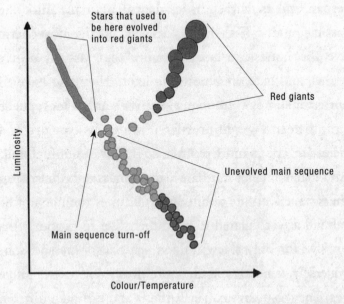

The Hertzsprung-Russell diagram: the main sequence turn-off tells the age of a globular cluster

back in time, and have remained undisturbed since then. So a 10 billion-year-old globular cluster will contain only red dwarf stars because all the yellow and blue-white stars have already died. Astronomers can estimate the age of a globular cluster using a Hertzsprung–Russell diagram. Named after the Dane Ejnar Hertzsprung and the American Henry Norris Russell, the Hertzsprung–Russell diagram (or HR diagram for short) was developed in about 1920 and is a way of understanding the lives of stars. The diagram is a plot of the surface temperature of a star against its brightness; every star in the Universe can be placed somewhere on it.

The surface temperature is linked to the colour of the star, and the brightness is linked to its surface area. Most stars fall on a diagonal band across the diagram from which the 'main sequence' stars take their name, massive blue stars to the left, red dwarf stars to the right and yellow medium-sized stars between the two. Only when a star begins to age and die does it move away from the main sequence, then becoming a red giant star. In the early lifetime of a globular cluster, blue stars would exist and would be becoming red giants. Later in the life of the cluster, there will be no blue stars left, and the smaller, yellow or red stars will be coming to the end of their lives on the main sequence. Hence, after plotting all the stars from a globular cluster onto an HR diagram, the point at which the main sequence turns off into the red giant region tells astronomers the cluster's age.

Using this method, astronomers have estimated that the Universe must be older than 12 billion years because that is the average age of the Galaxy's globular clusters. It means that at just 4.6 billion years old, our Solar System is a relative newcomer on the celestial scene.

It is not possible to do the H-R diagram analysis for the whole Galaxy because there are reservoirs of gas from which new stars are constantly being formed. Instead, astronomers have developed a different technique that relies on them scouring the Galaxy for the corpses of long-dead stars. These are known as 'white dwarfs' and are remarkable celestial objects. Each contains roughly the mass of the Sun but compressed into an object the size of the Earth, making it extremely dense. The white dwarf was once the nuclear furnace at the heart of a star at temperatures of tens of millions of degrees, so it takes a considerable amount of time to cool down. To use white dwarfs as age indicators, astronomers look for the coolest ones they can find and calculate the time it has taken for them to cool to their current temperature. The coolest ones observed have surface temperatures of a few thousand degrees and it is thus estimated that they are between 11 and 12 billion years old. Because white dwarfs are so small, most of those found so far are within our own Galaxy. Recently, however, the Hubble Space Telescope has found a number in a nearby globular cluster. These too provide ages of 11–12 billion years. It is hugely encouraging that the age estimates from both HR diagrams for globular clusters and cooling of white dwarfs converge on 12 billion years. This reliably sets a minimum age for the Galaxy; the challenge remains to deduce an accurate age for the entire Universe.

The Big Bang

The search for the age of the Universe received an enormous boost when Edwin Hubble discovered the expansion of the Universe in 1929 (see *How Big is the Universe?*), and it was confirmed that Albert Einstein's General Theory of Relativity could explain the observed behaviour. Einstein could have proposed the expanding Universe earlier, because his equations showed him that this movement was an intrinsic property of space. However, trapped by the old thinking that the Universe was infinite and static, Einstein had not believed his own work. Only Hubble's observations of redshift convinced him.

A young Belgian mathematician, Georges Lemaître, was bolder and began seriously investigating the properties of an expanding Universe when it was just a numerical possibility. Two years before Hubble made his observations known, Lemaître published a paper predicting that the Universe would be found to be expanding. It led him to the solution of Olbers' Paradox by making him contemplate the possibility of a creation event for the Universe. If the Universe was expanding today, then clearly it had to have been smaller in the past. To find out how much smaller, Lemaître used the mathematics of general relativity to model a reverse expansion, rather like a film running backwards. This allowed him to investigate the behaviour of the Universe at earlier and earlier times, when the celestial objects had been packed more closely together. He ran the cosmic clock all the way back until the entire Universe was compressed into a single, incredibly dense object that must have somehow exploded outwards.

At the time, scientists were fascinated with the recently discovered radioactive elements that seemed to spontaneously split apart, and Lemaître proposed that the Universe had been born when some 'primeval atom' spontaneously exploded. This was the beginning of what we now commonly call the Big Bang theory. Although scientists no longer think in terms of a primeval atom, much of cosmology today is concerned with understanding the nature of the Big Bang. To the astronomers in Lemaître's time, this was an extraordinary concept. The prevailing wisdom was that space had always existed and would always exist – which was convenient because they would not have to explain how the Universe came about in the first place. But, as we have seen, this blind assumption was eroded by Olbers' Paradox and then by Hubble's discovery of the expanding Universe. In the end, astronomers had no choice but to accept Lemaître's interpretation and, in the process, they realized that it gave them an extraordinarily simple way to estimate the age of the Universe.

The Hubble time

The theory of the expanding Universe throws up a value for the expansion rate of the Universe, called the 'Hubble constant'. Assuming that this has remained constant since the Big Bang event, cosmologists can use it to calculate the time it has taken for the Universe to expand to its present size. They call this the 'Hubble time', and using the presently accepted Hubble constant, it comes out at 13.7 billion years. But this is not quite the age of the Universe - it is an overestimate because of several factors that affect the expansion rate. One is the presence of matter. Matter produces gravity, which resists the expansion, and so astronomers have reasoned that the expansion rate will have slowed with time. Mathematical analysis of the expected slow-down shows that the actual age must be around two-thirds of the Hubble time, or just over 9 billion years. And that gives astronomers a huge problem because it conflicts with the ages of the globular clusters and the white dwarfs, which are both around 12 billion years. This conundrum is called the 'age crisis' in cosmology.

Perhaps the calculation of the Hubble constant from galactic redshifts and standard candle distance markers (see *How Big is the Universe?*) was flawed. Astronomers developed a new way of estimating it from studying the microwave background radiation – a perpetual sleet of microwaves that fills the Universe, believed to be the afterglow of the Big Bang. They analysed the pattern of hot and cool spots in this microwave background. Then, using general relativity and an estimate of how much matter and energy there is in the Universe, they deduced a Hubble constant – and this value was in excellent agreement with the traditional estimate.

A further refinement was then put into action, based on studies of patches in the microwave background created when the radiation collided with hot gas clouds deep in space. This value of the Hubble constant *also* agreed with the others. In one sense this is good news because all of the estimates are close to one another and reliably indicate a Hubble time of between 12 and 14 billion years, but the bad news is that it does not resolve the age crisis. Thankfully, an extraordinary solution appears to be at hand.

Cosmic acceleration

The question that cosmologists had to address was whether the presence of mass slows down the expansion over time as much as theory suggests. Recent redshift surveys have attempted to answer this. In the mid-1990s, two independent teams published their measurements and shocked everyone. Both teams found that the expansion rate is not decreasing at all. Somehow, the expansion of the Universe is accelerating. This finding went against all expectations and continues to fascinate and perplex in equal measure because no one knows what could be driving the acceleration. Some suspect it is a form of energy that Einstein originally toyed with during the development of general relativity, but this energy is difficult to conjure up from any modern theory. At present, there are no firm answers, and the mystery is encapsulated in the term 'dark energy' (see *What is Dark Energy?*).

One thing, however, is certain: if the Universe expanded more slowly in the past, then it has taken longer to reach its current size than the Hubble time suggests, and the age crisis is averted. By all modern estimates, the Universe is around 13.7 billion years old.

What are stars made from?
The cosmic recipe

Less than two centuries ago, what the stars were made of was the subject of myth and mystery. Some believed it would never be possible to find out the truth. But the development of spectral analysis and atomic physics in the 19th and 20th centuries has enabled revelations of stellar composition and cosmic abundance, of which earlier astronomers could only dream.

As surprising as it may seem, the chemicals we are most familiar with on Earth – oxygen in the air, carbon in living things, silicon in rocks – represent only a tiny fraction of the substances in the Universe as a whole. Across the entire cosmic landscape, it is hydrogen that turns out to be by far the most abundant chemical element. By mass, hydrogen accounts for 74 percent of all the atoms in the Universe; another 24 percent are helium atoms, and only the remaining two percent is made up of all the other chemical elements. In order to find out how astronomers have come by these figures, and how they have discovered the internal composition of stars, it is helpful to first delve a little into what an atom is like.

The nature of the atom

When, in 1869, Russian chemist Dmitri Mendeleev famously organized the chemical elements into the periodic table, which charts their similarities and differences, no one knew what gave the elements their chemical identities. That knowledge had to wait until after 1911, the year in which Ernest Rutherford and his team of physicists at Cambridge University's Cavendish Laboratory unravelled the structure of atoms.

Rutherford discovered that atoms have a highly dense, positively charged nucleus. He likened the nucleus within the atom to a gnat in London's Albert Hall, although most people today refer to the analogy as 'the fly in the cathedral'. It conveys the idea that most of an atom is empty space, with the majority of its mass concentrated into a tiny central region. The key to an atom's chemical identity is the composition of this nucleus. Inside it are minuscule subatomic particles called protons. Each one measures just one billionth of a micrometre across and carries an electrical charge. The number of protons determines how the atom behaves chemically. In the case of hydrogen, the nucleus contains a single proton, making it the simplest kind of atomic nucleus; oxygen has eight protons; carbon has 16; and so on across the periodic table.

The fact that the protons are all positively charged means they try to repel each other. To stop atoms spontaneously flying apart, nature has conspired to include other particles in the nucleus as well, to act like glue. Called neutrons, these subatomic particles are similar to protons but carry no electrical charge. They cannot exist for long outside an atomic nucleus - they decay in around 15 minutes if forced into isolation. Inside the nucleus, however, they are stable and they hold the protons together. The more protons in a nucleus, the more neutrons there need to be. For example, the second simplest atom, helium, contains two protons and two neutrons; lithium contains three protons and four neutrons; and so on. Unlike the number of protons, which defines its chemical properties, an element can contain different numbers of neutrons and still be the same element. These variants are known as 'isotopes' and can sometimes be radioactive (see *How Old is the Universe?*). Normally hydrogen has only a single proton and no neutron, but a version of hydrogen exists that also has a neutron; this 'heavy' hydrogen is an isotope known as 'deuterium'.

Being composed of differing numbers of protons and neutrons, it is obvious that different atoms will have different masses. The lightest element is hydrogen, and the heaviest naturally occurring element on Earth is uranium, with 92 protons and 146 neutrons. This difference in the mass of elements is the reason for the Earth's particular chemical identity: our smallish planet does not generate enough gravity to hold onto the lightest elements, such as hydrogen and helium. Mercury, Venus, Mars, and the moons that dot the Solar System, are also too small to retain hydrogen. Only when a planet reaches the mass of Jupiter or Saturn, approximately 300 times the mass of Earth, does its gravity become strong enough to retain every element.

Although the neutrons hold each atomic nucleus together, there is one thing they cannot do: cancel out the electrical charge of the protons. Every atomic nucleus is naturally positively charged. This is usually balanced by even tinier particles called electrons, which are negatively charged and encircle the nucleus. If there are ten protons in a nucleus, then ten electrons will orbit that nucleus, keeping the atom electrically neutral. To return to our analogy, the electrons will be microscopic specks spread throughout the whole cathedral, whizzing around the fly.

Electrons in atoms can absorb energy if the atom is bathed in radiation, and be boosted to a higher-energy orbit within the atom; when this takes place the atom is said to be excited. Very quickly the electron will drop back to a lower orbit and the atom returns to a more stable state, radiating energy as it does so. Electrons can even be temporarily stripped away from the atom completely if the absorbed energy is sufficient, leaving the atom positively charged. This is called ionization; the atom soon attracts another negatively charged electron to fill the gap, and radiates energy as the electron is captured into orbit. Strange as it may sound at first, this behaviour of atomic electrons has given astronomers a perfect tool for investigating the composition of celestial objects.

Seeing the light

In 1830, French philosopher Auguste Comte published his *Cours de philosophie positive*. In it was a remarkable claim that, at the time, must have seemed entirely reasonable. Comte stated that mankind would never know anything about the Sun's composition because it was too far away to analyse. By extension, the same went for the other celestial objects. The only exceptions to this rule were the occasional meteorites that fell from the sky, which could be collected for analysis. These were often made of iron, which at the time was taken to mean that perhaps metals were abundant throughout space.

Just three years after Comte's death in 1857, German scientists Gustav Kirchhoff and Robert Bunsen discovered a way to make the analysis of distant celestial objects common-place. It must have seemed like a dream come true and opened up the Universe for study in a way no one could have previously imagined. The pair was investigating the puzzling phenomenon that when sunlight is passed through a prism and split into its constituent colours, known as a spectrum, a pattern of vertical dark lines appears along the spectrum. It looks rather like a barcode superimposed on the otherwise continuous sweep of rainbow colours.

Around the same time, chemists were noticing that when they burnt a pure sample of a chemical element it turned the flame a characteristic colour. For example, sodium burns orange and lithium burns red. When this light is passed through a prism it resolves into a pattern of bright vertical lines, the brightest of which is the colour seen during the flame test; indeed for some elements, there is only a

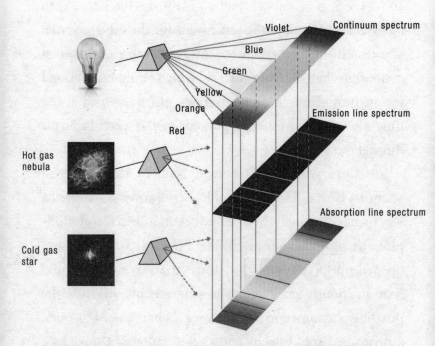

Spectral analysis: the pattern of spectral lines betrays a celestial object's chemical composition

single vertical line of colour. Wondering what the connection was between the dark lines in a continuous spectrum and the bright lines from flame tests, Kirchhoff began investigating using Bunsen's newly devised burner. He burnt chemicals, he split light through prisms, he investigated sunlight further, and eventually it dawned on him that elements only ever absorb the same wavelengths of light as they give out. In effect, they store the absorbed light energy for a short time and then re-emit it back into their surroundings. Sodium, for example, is only capable of giving out and absorbing yellow light with a wavelength of 589 nanometres; for lithium it is red light with a wavelength of 670 nanometres. With the hindsight of our knowledge of atomic structure and electron energy levels that was developed in the early 20th century, this all makes sense. But Kirchhoff made his analyses with no idea why elements showed characteristic spectra.

All elements have their specific wavelengths, like fingerprints, and whether they emit or absorb light depends entirely upon their physical condition. An incandescent vapour, such as one created when burning an element in a flame test, gives out that element's characteristic light. But a cooler vapour with a source of light behind it absorbs its characteristic wavelengths, creating dark lines in the light's spectrum. Astronomers swiftly understood that the Sun's surface was just such a light source and that it was surrounded by an atmosphere of cooler gases. Kirchhoff's work meant that they could deduce the chemical composition of the Sun and of other unreachable celestial objects. All they had to do was collect sufficient light to pass through a prism or some other device capable of splitting the light into a spectrum, and then measure the wavelengths of the dark lines. The chemical composition of the Universe opened up before their very eyes.

Cosmic composition

Early spectral analysis made it look as if the stars were largely composed of metallic elements, mirroring the meteorites that fell to Earth. But the presence, or not, of a specific dark absorption line was later understood to be dependent on the ionization of the atoms, that is whether one or more electrons had been stripped from the atom. It meant that not every element present would necessarily show up in an absorption spectrum. After decades of laboratory work to understand the effects of ionization, correction factors have allowed us to arrive at the correct cosmic abundance of elements: 74 percent hydrogen, 24 percent helium, and two percent all the other chemical elements.

This does not mean, however, that these proportions have been fixed since the moment of the Big Bang. The stars themselves are chemical factories, converting the Universe's vast stocks of interstellar hydrogen into the heavier chemical elements. This fearsome activity, known as 'nuclear fusion', takes place at the core of each and every star, where the temperatures reach millions of degrees. It was in the early years of the 20th century that French physicist Jean-Baptiste Perrin suggested that the nuclear fusion of hydrogen was at work in the Sun, but it wasn't until 1938 that scientists understood the details of the mechanism.

The Sun's nuclear reactor

In the Sun's core, where the temperature reaches 15 million degrees and the gas is squeezed to about 150 times the density of water, nuclei of hydrogen are being fused into nuclei of helium, releasing vast amounts of energy. Four hydrogen nuclei, that is four protons, are converted into a single helium nucleus – two protons and two neutrons bound tightly together. Even at the density of the Sun's core, it is unlikely that four nuclei will collide with one another simultaneously; instead, a helium nucleus is built up in a sequence of collisions, during which two out of four protons convert to neutrons. Energy is released at all stages as mass is lost, thus causing the star to radiate heat and light.

About 4 million tonnes of matter are transformed into energy every second inside the Sun – enough mass to build more than ten Empire State Buildings. This energy pushes its way outwards, taking hundreds of thousands of years to reach the surface, where the temperature is a mere 6000 degrees and the density is only a tenth of a percent of the density of Earth's air at sea level. From here, the energy shoots off into space as electromagnetic radiation. If it happens to be heading for Earth, its journey will end just eight minutes later in our skies. It is remarkable to think that the warm sunlight on your face was generated in the heart of the Sun, hundreds of thousands of years ago.

Investigating the way in which stars build other, heavier elements through nuclear fusion and how that alters a star's composition is not easy, since those changes are locked into the core of the star. Only in the star's death throes does it effectively dissect itself for the whole Universe to see – and on Earth there are plenty of astronomers poised with their telescopes.

The death of stars

Broadly speaking, stars die in one of two ways, depending on how much mass they contain. The first scenario applies to the low-mass stars, which contain less than eight times the Sun's mass. As with all stars, energy is generated only in the core and as more and more of the hydrogen in the core is converted into helium, so the process falters. Although the star's surrounding layers contain more than half the mass of the star, this matter cannot penetrate the core to replenish the waning stocks, and so, as hydrogen fusion in the core slows down, the outpouring of energy dwindles and the star begins to contract. This squeezing drives up the temperature and ignites hydrogen fusion in a shell around the now largely inert core. This shell releases a new rush of energy that pushes the upper layers of the star outwards, bloating it from a diameter of roughly a million kilometres to hundreds of millions of kilometres. The outer layers cool and the star becomes a 'red giant'. When this happens to the Sun (in five billion years time) it will engulf Mercury, Venus and Earth.

Inside the red giant star, the core continues to shrink and heat up. When its temperature reaches 100 million degrees and its density approaches 1000 times that of water, the helium begins to fuse into carbon, and into oxygen if the temperature is high enough. This activity lasts for between 10 million to 100 million years and then the helium runs out and the star contracts again. The shell of hydrogen fusion continues around the core, creating enough energy to lift the outer layers still further, blowing them off into space to create a nebula of glowing gases. William Herschel, the 18th century astronomer who had recently discovered Uranus, thought these nebulae looked rather similar, so he misnamed them 'planetary' nebulae and the name has stuck.

The more layers the dying star sheds, the closer to the core astronomers can peer. Spectroscopic studies of planetary nebulae do indeed show enrichments of heavy elements, but still most of these remain locked in the stellar core which sits at the heart of the nebula. As the surrounding layers lift off into space, the core can eventually be seen as a sphere of compressed gas, about the size of the Earth but 200,000 times more dense, with a temperature around 500,000 degrees. These are 'white dwarf' stars that astronomers use to age the Milky Way (see *How Old is the Universe?*). They are composed of almost pure carbon and oxygen, with small quantities of the other chemical elements.

By contrast, the second scenario applies to high-mass stars, with more than eight times the mass of the Sun. The greater mass generates even higher temperatures in the core following the cessation of hydrogen fusion. These temperatures allow the star to build ever-heavier chemical elements through fusion, and spur the star on to expand to gargantuan proportions. If one of these red 'supergiants' were to magically replace the Sun, it would reach to the planet Jupiter. These stars suffer a more spectacular fate than their low-mass cousins.

The evolution of a red supergiant follows the same pattern as a red giant up until the helium fusion stage. Then, because of the extra mass pushing down on the core, it experiences further bouts of re-ignition, allowing it to fuse heavier and heavier chemical elements. The star develops an onion-like structure of layers around the core, each with different nuclear fusion reactions taking place: from the outside layers to the inner layers, there is fusion of hydrogen to helium, helium to carbon and oxygen, carbon and oxygen to neon and magnesium, and neon and magnesium to silicon and sulphur. At the very centre of a mature red supergiant, silicon and sulphur fuse into iron and nickel; and this is the death knell for the star.

All other fusion processes in the star have released energy, but to fuse iron and nickel requires energy to be put into the reaction. Since the energy generated inside the star flows outwards towards the less dense regions – like water running downhill – there is no available energy in the core to fuse the iron and nickel, and the generation of further energy stops. According to computer simulations, it takes just 24 hours to build up an iron–nickel core containing almost one and a half times the mass of the Sun. Having achieved this mass (the Chandrasekhar limit, see *How Big is the Universe?*), gravity overwhelms the forces holding the atoms apart, and the core rapidly collapses into a ball of tightly packed neutrons just 10 to 20 kilometres (6 to 12 miles) across. When it does this, the upper layers come crashing down on the core, initiating a tremendous explosion called a type II supernova. The energy released in the supernova drives a final giant burst of nuclear fusion that builds all the chemical elements heavier than iron and nickel, including radioactive elements. In the explosion, these newly minted elements are scattered throughout space.

Astronomers measuring supernova remnants today find some that are still expanding into space at a few thousand kilometres per second. The gases glow with their characteristic colours even after the initial energy of the explosion has been dissipated, because the radioactive elements in their midst decay, providing more energy to keep the gases excited. Eventually, the debris from both planetary nebulae and supernova remnants merge into the general reservoir of interstellar gas and dust, where they are subsumed into newborn stars. Overall, this grand process of stellar life and death has the effect of enriching the chemical composition of the Universe with heavier elements.

So the answer to the question 'What are stars made from?' is that stars are made from the debris of the former celestial generations, and in this way each new generation contains a greater proportion of heavy elements than the last. Even though these account for only two percent of atoms across the Universe, they have a profound effect because they allow planets like the Earth to form, and life to emerge.

How did the Earth form?
The birth of the planet we call home

With its profusion of landscapes and climates, animals and people, our Earth certainly seems to be something special and deserving of the name 'cradle of humankind'. To planetary astronomers, however, the Earth and the other planets of the Solar System are cosmic flotsam, debris left over from the formation of the Sun, itself just an average star in a commonplace galaxy.

The Earth and the rest of the Solar System formed during a hellish maelstrom around 4.6 billion years ago. Such activity is driven by gravity and continues in other parts of the Universe today, forming new stars and planets, but it is all hidden away inside dusty cocoons that absorb ordinary wavelengths of light. To peer inside these celestial construction sites, astronomers have to use other wavelengths, such as infrared or microwaves. The goal of this research is simple: to see what is happening in these regions of activity and relate it to what must have occurred during the formation of our own Solar System. By doing this, the astronomers hope to solve a number of mysteries.

The first of these unknowns is why the planets of the Solar System all orbit in more or less the same plane. Presumably, there must have been some natural process that prevented the planets from assuming random orientations and corralled them into nested orbits. The second is, why there is such an abrupt change in the composition of the planets from the inner to the outer Solar System. Some process segregated the raw material to produce relatively small rocky worlds in the inner Solar System and giant gaseous planets further out. To answer these questions, astronomers must identify and observe sites of star and planet formation elsewhere in the Galaxy.

Giant molecular clouds

Stars and planets account for 85 percent of all the atoms in the Galaxy, leaving around 15 percent available in interstellar regions where new stars and planets form. This material floats through space creating a tenuous mist of gas and dust, of which hydrogen is the most abundant constituent, and the 'dust' consists of particles of heavier elements from planetary nebulae and supernovae (see *What Are Stars Made From?*).

In space there are typically only 100,000 hydrogen atoms in every cubic metre (at sea level on Earth the atmosphere contains around 200 million trillion atoms). While floating through the Galaxy, the hydrogen atoms tend to pair up into molecules, and the molecules gather together into enormous clouds. Astronomers estimate that there are about 4000 of these giant molecular clouds in the Milky Way Galaxy. Each one is typically between 150 and 250 light years in diameter and contains enough gas to make up to ten million stars like the Sun. As the molecular gas falls together it sets the cloud rotating, a fact that will become very important for the subsequent formation of stars and planets.

Within each giant molecular cloud there are a multitude of slightly denser clumps, each measuring between a quarter and a third of a light year across. These clumps are the seeds from which stars grow – or, more accurately, shrink. A clump generates gravity and pulls itself together in a slow process that takes anything from a million to ten million years. Shockwaves from nearby supernovae explosions may speed things up by compressing the gas and also triggering further clumps to form. As a clump shrinks, so the gas and dust inside becomes denser. Dust accounts for only one to ten percent of a clump's mass, yet this is enough to eventually block our view of the newly forming star at ordinary wavelengths of light. As astronomers look into the dusty cocoons with infrared telescopes they see something remarkable: each collapsing clump transforms itself from a nearly spherical shape into a disc surrounding the nascent star. These flat discs can be anything from 100 to 1000 times larger than the Earth's orbit, and it is in these that planets form.

Solving the mysteries

Planet-forming discs are flat because of centrifugal force. This is not a force based upon an intrinsic property, such as electrical charge, but is instead created when an object begins to spin. The faster something rotates, the greater the centrifugal force it generates, pushing objects outwards: so a ball placed on a merry-go-round will roll towards the outer edge as the merry-go-round begins to spin. It is also the reason a car will leave the road if the driver takes a bend too fast.

Consider the clump of gas and dust that formed our Solar System. As the spherical clump contracted under its own gravity, it naturally spun faster, in the same way that a spinning ice dancer speeds up when she pulls in her arms. This would have boosted the centrifugal force felt in the equatorial plane of the sphere, opposing the pull of gravity on the gas and dust and slowing the collapse in this region. Away from the equator, where the effects of centrifugal force were not so great, the gas and dust would have fallen inwards more or less freely, so the overall result was that the cloud pancaked into a flat disc. Within this disc the planets subsequently materialized, and therefore they occupy almost identical orbital planes rather than whizzing around the Sun in random orientations. This provides the solution to the first mystery.

The second mystery was why the composition of our Solar System planets changes so drastically from the inner regions to the outer regions. In the inner Solar System, the planets are like the Earth – small, rocky, and with thin atmospheres – whereas in the outer Solar System they are more like Jupiter – large, gaseous, and with thick atmospheres. This difference comes about because, as the newly forming Sun shrank to its modern size and density, it released energy; nowhere near as much energy as would be released later when nuclear fusion ignited in its core, but nevertheless enough to heat the surrounding disc. The heat prevented certain chemicals from forming by driving particular atoms into frantic motion. Other atoms and molecules were not so affected by the heat. They bumped into one another and bound together, creating dust of different chemical compositions. The chemicals that formed depended on the temperature, which in turn depended on the distance from the forming Sun. Near the Sun, in what would eventually be Mercury's orbit, the temperature would have been several thousands of degrees and only metallic and some silicate atoms could condense into dust (hence Mercury's metallic core takes up a large proportion of is volume); other chemicals would have been vaporized back into gas by the heat from the young Sun. In the somewhat lower temperatures near the Earth's eventual location, more silicates would have been able to condense (giving us a smaller metallic core in comparison with the rest of our planet). Further out still, the lower temperatures enabled other elements to form dust.

The snow line

A distinct boundary in the planet-forming disc occurred at five times the Earth's distance from the Sun. Called the snow line, it is where Jupiter orbits today. At the snow line, when the planetary material was condensing, the temperature would have been around 90 K, low enough for molecules such as water, ammonia and methane to form ice. (Astronomers measure temperature in kelvin, K. Zero on the kelvin scale is the temperature at which atoms and molecules all cease to transfer energy between one another, known as 'absolute zero', this is the equivalent of approximately −273 degrees Celsius.) At 40 times the Earth's distance from the Sun, roughly where Pluto orbits, the temperature would have been just 20 K and almost every chemical element could condense; the only exceptions were hydrogen and helium, which remained in a gaseous state. Consequently the dust grains developed different compositions throughout the planet-forming disc, leading to the variety of planets. The much larger size of the outer planets can be explained by the much greater reservoir of matter beyond the snowline.

In the inner Solar System, the dust gradually accumulated into objects resembling small asteroids; these 'planetesimals' would have populated the early Solar System in vast numbers. To build Mercury, Venus, Earth and Mars would have required ten billion or more planetesimals of 10 kilometres (6 miles) in diameter. These lumps of rock continued to grow by basically colliding and sticking together, but the process was subtler than it may at first sound.

Head-on collisions were no use because they released too much energy and would have shattered the planetesimals, blasting the debris into space. In any case such energetic clashes would have been rare because the planetesimals were all rotating in the same direction. Sometimes an impact was just enough to melt the planetesimals together, at other times although it broke them into fragments, the pieces remained together and continued to orbit as a pile of rubble.

These close encounters were repeated along entirely random lines until eventually some larger planetesimals began generating enough gravity to pull smaller ones onto themselves. Throughout the disc, these major planetesimals began to outpace their lesser companions and, the bigger they became, the more efficient they grew at drawing in smaller bodies. Astronomers call these planetesimals 'oligarchs', because they controlled their surroundings. Essentially they were small rocky planets, each containing between the mass of the Moon and Mars; computer simulations show that 20 to 30 of them must have ultimately smashed together to build the Solar System's four terrestrial planets of today.

Gas giants and planets beyond

The Solar System's gas giants – Jupiter, Saturn, Uranus and Neptune – were probably formed in a similar way to the inner planets but from bigger oligarchs, bolstered by the astronomical ices. Once the forming Jupiter and Saturn reached between three and five times the mass of the Earth, they generated so much gravity that they began pulling in gas from their surroundings and this gave them their thick atmospheres that mirror the cosmic abundance of elements. Uranus and Neptune formed in a similar fashion although, being less massive, they were not so good at attracting the hydrogen and helium, so they display a greater proportion of astronomical ices in their atmospheres.

A number of astronomers have put forward the alternative suggestion that the gas giants formed in the same way as stars do. In this scenario, a region of the disc around the Sun would have reached a critical density and gravity simply pulled it all together. There were no oligarchs building up and colliding, just a sudden collapse of gaseous matter into a giant planet. At present there is no way to determine which of these gas giant formation scenarios actually took place in the Solar System. Both theories correctly predict that the giant planets surround themselves with their own mini-discs, which subsequently coalesce into extensive moon systems.

At the distance of Pluto, the density of matter orbiting the young Sun was thinner, so the bodies that formed there were consequently smaller. Pluto itself is only two-thirds the size of Earth's Moon, and has an orbital plane significantly inclined to that of the other planets. In 2006, these factors, together with the discovery of a number of other Pluto-like objects in the outer reaches of the Solar System, led to the International Astronomical Union voting to downgrade Pluto from the status of planet to dwarf planet. The newly observed bodies included the icy celestial objects Haumea and Makemake, but it was the body catalogued as 2003 UB313 that really triggered the debate. Observations showed that it was at least the size of Pluto, probably bigger. Hence, astronomers faced a choice: downgrade Pluto or name 2003 UB313 as the tenth planet in the Solar System. A heated discussion ensued, during which 2003 UB313 was nicknamed Xena, after the television heroine *Xena: Warrior Princess*. Eventually, the decision was made: 2003 UB313 was not a planet and as a result Pluto was downgraded. Xena's name was changed to Eris – rather appropriately – after the Greek goddess of strife and discord.

There are undoubtedly many more dwarf planets yet to be found in the Solar System. Astronomers estimate that there could be hundreds or even thousands of them beyond Pluto, and possibly a few fully-fledged planets. In fact, according to computer simulations, a whole second Solar System's worth of planets may be lurking at thousands of times the distance of Earth from the Sun. These could turn out to be as big as Mars or even Earth – not formed in situ, but thrown there by the gravity of the gas giant planets. If a planetesimal were travelling sufficiently fast near a giant planet, the giant's gravity would be unable to pull it into a collision. As the smaller object sped by, the near miss could result in it gaining considerable speed and being boosted into a larger orbit. In this way, Jupiter could have scattered rocky planets out to between 25 and 250 times further from the Sun than Pluto.

At such a distance from the Sun, the scattered planets would be very faint indeed and so extremely difficult to spot. Add to this that Jupiter could have catapulted them into any orbital plane, and the only way astronomers will be able to search for them is to trawl the whole sky with a powerful telescope. There are a number of such instruments on the drawing board at the moment, all due to begin searching within ten years.

The heavy bombardment

By 4.6 billion years ago the Solar System looked almost as it does now; the familiar planets and their moons had formed. The space between the planets, however, remained home to countless, smaller leftovers. These tiny objects ranged from pebbles and rocks to planetesimals that had so far escaped the planets' gravitational clutches. Jupiter's gravity trapped many of them between itself and Mars to form the asteroid belt, but most whizzed about the Solar System. During the next 700 million years these celestial vagabonds collided with the planets and their moons, blasting out craters of all sizes. Old planetary surfaces are easily identified today because of their heavily pockmarked appearance: our Moon being the classic example. Its scarred face has taught astronomers much of what they know about this last phase of planet formation, referred to as the 'heavy bombardment' period. On Earth, the early craters have been eroded away; today, less than 200 craters are known, and all of them are from comparatively recent impacts.

The position of the Earth inside the snowline meant that it was formed without any water; the heat from the young Sun would have vaporized any water molecules that formed. This is another puzzle that astronomers have had to address: how we came to have oceans. The 'late bombardment' suggests a way; planetesimals that formed in the outer Solar System, incorporating water and other ices, rained down on Earth and the other planets of the inner Solar System, supplying them with water and other volatile substances that they lacked. On Earth this material was swiftly transformed into life (see *Are We Made from Stardust?*).

During the bombardment period, many of the planetesimal remnants would have been ejected from the solar vicinity by near misses with Jupiter, in the same way that Jupiter is thought to have scattered planets. Because they were much smaller than planets, Jupiter could have lofted them much further, throwing a trillion or more of them into incredibly large orbits, reaching 10,000 to 100,000 times the Earth's distance from the Sun. This distant collection is called the 'Oort Cloud'. Occasionally one of its members returns to the inner Solar System, and we call these distant visitors 'comets'. Being composed of ice they begin to melt when they approach the Sun, and leave a trail of gases in space that we see illuminated as the comet's tail. Dust released from the melting ice litters interplanetary space, and if the dust falls into the Earth's atmosphere it burns up and creates meteor showers, or shooting stars. Now and then, if a conglomerate of dust is large enough, it might not burn up entirely but plummet all the way to strike the ground as a meteorite. Fragments chipped from asteroids can also fall as meteorites.

And what of the future?

After some 700 million years the bombardment petered out, and the formation of the Earth and the rest of the Solar System as we know it was complete. But even now we catch a glimpse of what the late bombardment must have been like. In 1993, fragments of a comet called Shoemaker-Levy 9 repeatedly struck Jupiter. The 21 fragments, some of which were 2 kilometres (1.2 miles) wide, slammed into the giant planet over a six-day period. Each struck with a speed of approximately 60 kilometres per second (130,000 miles per hour), causing tremendous explosions that left plumes of debris, some larger than the Earth, visible in the planet's atmosphere for weeks afterwards.

Orbiting spacecraft have shown that comets the size of a two-storey house regularly approach the Earth, but fortunately they break up in the atmosphere and present no danger. However, occasionally a large meteorite hits the ground: notably the strike on the Tunguska region of Siberia in 1908, which razed an area of uninhabited forest the size of a modern city such as London.

Today, a constant watch is kept for threatening asteroids. Although most asteroids are safely corralled in the main belt between Mars and Jupiter, an increasing number of 'near-Earth objects' are being discovered. Astronomers track almost 800 nearby objects with diameters of greater than 1 kilometre (0.6 miles). Of these, none presently pose a threat to Earth although calculations suggest that, on average, something of this size hits us every half a million years. There may be many thousands of nearby asteroids less than 1 kilometre across, of which only a small percentage is currently being tracked. One, catalogued as 2007 VK184 and about 130 metres (520 feet) wide, is already known to present a slight danger. While tiny in asteroid terms, nevertheless should it hit the Earth it would release energy equivalent to 10,000 Hiroshima atomic bombs. It currently has a 1-in-3000 chance of hitting Earth in 50 years' time, but this probability is expected to decrease to zero as further observations allow astronomers to refine its orbit. Better and better survey equipment is constantly being developed to reveal and track such near-Earth objects.

Inevitably, one day an asteroid will be discovered to be on a collision course with Earth. As soon as its path can be reliably predicted, plans will be put into action to attempt to deflect it, and astronomers have a number of methods in mind for doing this. Blowing the asteroid to pieces with nuclear weapons is not the answer because this would not alter the orbit of the fragments – instead of coming towards us as cannonball, it would be transformed into buckshot. Exploding nuclear weapons some way above the asteroid's surface might be a more successful, if tricky option; some of the rocks would be melted and would release gases creating a natural 'rocket' engine to nudge the asteroid onto a safer orbit. It seems somewhat ironic that the very objects that brought water to Earth and made life possible now threaten to destroy it. When it comes to safeguarding the Earth against dangerous impacts, it is literally a case of 'watch this space'.

Why do the planets stay in orbit?

And why the Moon doesn't fall down

The question is a simple one. The answer is not as simple as the question, and the search for it led to the defining moment of science: the moment that precipitated the scientific revolution and the Age of Enlightenment; the moment when natural philosophy was placed on its course to become modern science, with mathematics as its language.

The reason why the planets stay in orbit is contained within Isaac Newton's theory of gravity, as detailed in his groundbreaking book, *Philosophiae Naturalis Principia Mathematica*, usually referred to as the *Principia*. In Newton's time, the late 17th century, the question was posed as 'Why does the Moon stay in orbit and not come crashing down on us?' but the solution is exactly the same; it is just a question of scale. The Moon orbits the Earth while the planets, including the Earth, orbit the Sun. The answer rests on a mathematical understanding of gravity.

The central concept of the *Principia* is 'universal gravitation'. This states that everything with mass generates gravity: the Earth, the Moon, the Sun, all the planets and the moons, all the stars – everything. The amount of gravity generated by a celestial object is proportional to the mass it contains and the resulting gravity affects anything else with mass that is nearby; in other words, everything pulls everything else. Newton's Theory of Universal Gravitation broke the mould of human thinking. It was partly based upon the pioneering work of Johannes Kepler, who had mathematically described planetary motion several decades earlier.

How the planets move

The first of Kepler's three laws of planetary motion states that each planet moves in an elliptical orbit, with the Sun at one focus. (An ellipse has two focal points and the further apart these foci are, the more elongated the ellipse.) Previously, planets had been assumed to move in circles, but Kepler deduced from analysing the detailed observation data of the Danish astronomer Tycho Brahe that all the planets follow their individual elliptical orbits.

The second law he deduced is that, as a celestial object follows its orbit, it sweeps out equal areas of the ellipse in equal times. To visualize this, imagine a long extensible tether between a planet and the Sun. As the planet moves a small distance through its orbit, so the tether pivots at

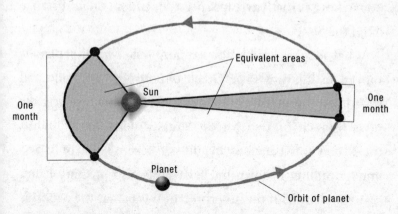

Kepler's second law: this law of planetary motion states that equal areas are swept out in equal times during the course of an elliptical orbit

the Sun, sweeping out a triangular area. When the planet is far away from the Sun, the tether is longer and the planet does not have to move far to sweep out a large area. Conversely, when the planet is close to the Sun, the tether is much shorter and the planet has to move much faster to sweep out the equivalent area in the same time. Hence, what the law is saying is that when a planet is far from the Sun it travels slowly and when it is closer to the Sun it travels faster. This was an important clue because it implied that whatever force was moving the planet weakened with distance.

The third law follows on from the second, expressing as an equation the link between the size of the planet's orbit and the time it takes to complete that orbit. In essence, it quantifies a planet's average speed dependent on its distance from the Sun.

What Kepler could not explain was *why* the planets moved in this way. The Greek philosopher Aristotle had stated that everything finds its place in the Universe because of a balance of two forces: 'gravity' and 'levity'. Thus the Moon was balanced in the sky because not only was gravity pulling it down but levity was also pushing it up. The problem with this interpretation was that the Moon is not balanced in the sky; it is moving in orbit around the Earth.

Apples and cannon balls

Newton showed that the Moon does indeed stay in orbit because of the interplay of two forces, but not ones that are in opposition to one another; instead, they must act at right angles. In fact English experimenter Robert Hooke was probably the first person to realize this, but he could not draw together the mathematics to prove it. This took the mathematical ingenuity of Newton. It is popularly thought that Newton took his inspiration from watching an apple fall but this story is apocryphal, or certainly not verifiable, as Newton never wrote about such an incident himself – but what he did write about were cannon balls.

Newton asked his readers to visualize a cannon atop an incredibly tall tower, pointing horizontally and firing its projectile. If we ignore the effects of air resistance, the cannon ball will zoom off parallel to the ground. However, the gravity of the Earth will immediately begin to pull it downwards, eventually dragging it all the way to the ground. The greater the explosive charge, the faster the projectile will be ejected and the further it will travel before gravity pulls it down. Now imagine the cannon loaded with sufficient explosive to eject the cannon ball so fast that, by the time it has started to fall, the curvature of the Earth has

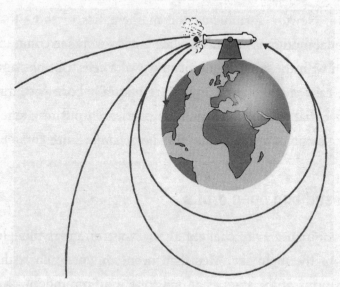

Cannonballs and orbits: the different amount of explosive charge determines which path the cannon ball takes

resulted in the ground beneath dropping away and so the cannon ball finds itself still at the same altitude above the Earth. Remember that we are neglecting air resistance, so the projectile is still travelling at the same speed as when it left the cannon, and the whole situation starts again. Every time the cannonball drops a little, so the curvature of the Earth compensates, allowing the projectile to continue around the Earth – forever. In effect, it has been placed into orbit.

This theoretical analogy gives us the solution to the question of what stops the Moon crashing into the Earth. The Moon is falling towards us, but also travelling along so fast that it 'overshoots' the Earth and continues in a circular path. Newton formulated the mathematics to show that orbital motion is produced when gravity works in conjunction with the tangential movement of a celestial object. In the case of the planets, they were naturally born with this tangential motion because they condensed out of a spinning cloud of dust and gas (see *How Did the Earth Form?*).

Closed and open orbits

Newtonian gravity changed the way astronomers thought about the night sky. Most had been content to chart the positions of the stars as an aid to navigation; indeed, this was considered the principal use of astronomy. After Newton's work, however, they could understand the motion of the celestial objects and, more importantly, predict their future movements. The dates of future eclipses, the return of comets, the conjunctions of the planets – all were prescribed by Newton's theory. It also showed that there are four possible orbital shapes: circles, ellipses, parabolae and hyperbolae. These can be

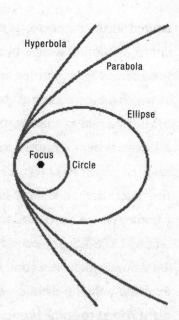

Orbital shapes: orbits come in four basic types

understood in terms of the cannon on the tower. First is the circular orbit, produced when just enough explosive is used to stop the cannon ball falling to Earth. Next is the elliptical orbit, produced by increasing the amount of explosives so that the cannon ball increases its altitude above the Earth before falling back and starting the circuit again. The more explosive used, the more elongated the orbit becomes until, eventually, it opens up and the projectile escapes altogether from the Earth's gravitational attraction. When just enough explosive is used to break the cannon ball free, it follows a curve called a parabola; as even more explosive is used, the curve opens up wider, becoming a hyperbolic shape. In both cases, the projectile never returns to Earth. The velocity at which a projectile must be fired to place it on a parabolic orbit is known as the 'escape velocity'. For Earth, this is 11 kilometres per second (7 miles per second). It varies from celestial object to object, depending upon how much mass the object contains. For Mars it is 5 kilometres per second (3 miles per second), and for Jupiter it is 60 kilometres per second (37 miles per second). The value for the Moon is just 2.4 kilometres per second (1.5 miles per second), which explains why the Apollo astronauts did not need another giant rocket to come home.

When a celestial object follows a closed orbit in space, where there is no air resistance, it circulates time and time again, and ellipses and circles are the observed orbital shapes for the planets, moons and asteroids in the Solar System. Some comets, such as Halley's famous one that we see every 76 years, also follow closed orbits. Stars follow closed orbits around the centre of their galaxy and even the mighty galaxies follow closed orbits around the centres of galaxy clusters.

By contrast, an open orbit is a one-time deal. A celestial object on an open orbit makes a fleeting visit and then disappears off into deep space never to be seen again. Many comets follow open orbits, and unmanned space probes have been directed into open orbits to view the surface of one planet before coasting onwards to the next.

Back on Earth

As important as the understanding of orbits proved to be, Newton's theory applied to a breathtaking range of other phenomena, not all of them celestial. His work gave the natural philosophers of the day a way of estimating the mass of the planets and the Sun, and a means of explaining why the Earth and other planets bulged at the equator. It gave those more minded of engineering problems a method of calculating the movement of falling objects on Earth, and, not least in the 17th century, of predicting the trajectory of projectiles fired from cannons. All motion, it seemed, could be understood in Newtonian terms. The Universe behaved as a clockwork mechanism, unremittingly following the laws that he had deduced.

Unsurprisingly, Newton's work was heralded as 'the system of the world'; a phrase for what we would now refer to as 'the theory of everything'. Over the course of the next few centuries, scientists came to realize how much else there was still to understand in the physical world: electricity and magnetism, nuclear forces, and relativity and quantum effects. But at the time, Newton's work was a triumph, and one of its victories was providing the explanation of the tides. The tidal changes were an all-important phenomenon for a sea-faring nation at that time, but their cause was a mystery until Newton proved in the *Principia* that they were due to the gravitational attraction of the Moon and the Sun on the oceans.

Consider the Moon's gravity pulling on the Earth: the Moon's tug is stronger on the side of our planet facing the Moon than on the opposite side, because gravity weakens with distance. This pulls the Earth into an elongated shape, which we experience as tides because the water is much freer to move than the rocks. The rocks beneath our feet do move too, albeit by less than a metre per day. The Sun's gravity also has a tidal effect and it is the interplay between the solar and lunar gravitational forces that gives the different height of the tides at different times of year. When the Sun, Earth and Moon all fall into approximately a straight line, we get a large tide, called a spring tide; when the configuration is perpendicular, we get the low, neap tides.

In the same way, the Earth's gravity simultaneously deforms the Moon and, because Earth is larger and more massive, the lunar tide is correspondingly larger, amounting to an elongation of the Moon by many metres. These changes to the spherical shape of the Earth and the Moon have an inertial effect, making it harder for each to spin. This constantly saps their rotational energy, and in the case of the Earth it causes the length of the day to slowly but perceptibly increase. This is one of the reasons why an extra second must be occasionally added to the midnight chimes at New Year. Called a 'leap second', it prevents the day-lengthening accumulating and causing the time of day to fall out of step with the Sun's position. The slowing of the rotation of the Moon is more profound; over the billions of years since its formation, the Moon's rotation has slowed so much that it is now locked into rotating just once every orbit. This is why the Moon constantly presents the same face to Earth.

Astronomers see tidal forces at play wherever two large objects are in orbit around one another. Because of its large size, Jupiter creates enormous tidal forces on its collection of moons. The one that suffers the most is the innermost moon, Io. This world is just a little bigger than our Moon at 3640 kilometres (2261 miles) across, and is the most volcanically active place in the Solar System. Io's volcanoes are constantly erupting, spewing sulphurous lava onto the moon's surface. The energy to drive this extraordinary activity comes from the tidal force that Io feels from Jupiter. This periodically squeezes Io, producing heat that melts the interior, driving the eruptions. Further out from Io is the moon Europa; because of the larger distance, the tidal force is less extreme and there is no spectacular volcanism. There is, however, strong evidence that below Europa's icy crust is a global ocean of water, which is kept liquid by the squeezing of the tidal force. This ocean may be anything from 10 to 100 kilometres (6 to 60 miles) deep – if so, there is more water on Europa than there is on Earth.

On a much larger scale, tidal forces are responsible for stretching whole galaxies. If two galaxies approach each other on a collision course, the strength of gravity acting on the near side of each galaxy is stronger than that acting on the far side. So the near sides accelerate faster than the far sides, elongating the galaxies as they plunge into one another. In the extreme, when matter falls into a black hole, it is elongated so much that it is literally pulled apart in an event known, somewhat tongue-in-cheek, as 'spaghettification' (see *What is a Black Hole?*).

Wobbling stars

Astronomers continue to exploit Newton's gravitational theory for further discoveries about the Universe. Over the last two decades, it has allowed them to find more than 400 planets orbiting other stars; they have only actually seen a few of these planets, but their presence is certain because the stars are 'wobbling'. We are used to the idea of a planet orbiting a star but that is only half the story; just as the star pulls the planet into an orbit, so the planet pulls back on the star. But because the star is so much more massive, the planet cannot cause the star to move through a large orbit; instead, it makes the star wobble. Take for example, our largest planet Jupiter. The Sun pulls it around its 750-million-kilometre (466-million-mile) orbit in twelve Earth years. In the same time, all Jupiter can manage is to force the Sun to pirouette about a point approximately 50,000 kilometres (31,000 miles) above the fiery surface. Such a pirouette is the periodic wobble that astronomers look for when searching other stars for planets. The size of the star's movement and the time it takes to complete one wobble allows astronomers to compute the mass and orbital radius of the unseen planet. The surprise is that instead of finding slow pirouettes like the Sun's that take years, most of the wobbling stars found so far have shown that their planets are large like Jupiter but complete their orbits in a matter of days. This indicates that the planets orbit very close to their parent stars (see *Are There Other Intelligent Beings?*). As technology improves and data is collected over longer timescales, astronomers expect to find planetary systems more nearly resembling our own.

A dilemma

For all its success, there was a paradox at the heart of Newton's work. Newton knew it was there and even used it later in his life to defend himself when he was accused of being an atheist. The paradox lay in the concept of 'universal' gravitation: the idea that everything in the Universe generated gravity, including the stars. If the stars were all pulling on one another, Newton could not understand why there was not general collapse. Observation suggested that the stars were in the same positions that they had been since the earliest ancient records, forming the same constellations that the Babylonians and the Greeks had seen, so naturally the assumption was that the Universe was static. Yet, paradoxically, the otherwise successful Theory of Universal Gravitation implied that it should be collapsing inwards.

Newton's religious beliefs were being questioned because his theory seemed to dispense with the need for God to move the heavenly objects about. To get around both the scientific conundrum and the religious criticism, Newton stated that it must be the hand of God preventing the Universe from collapsing. The real answer, not determined until centuries later, is that the stars are in orbit around the centre of the Milky Way, so they are supported by their own orbital motion in the same way as the planets in the Solar System.

Nowhere in the *Principia* did Newton explain the nature of gravity; his success was to describe it mathematically. Subsequent natural philosophers and scientists grappled with the fundamental origin of gravity, though none came close to any discovery and the world had to wait until the second decade of the 20th century to receive a mind-bending answer from Albert Einstein with his General Theory of Relativity.

Was Einstein right?
Gravitational force versu
space–time warp

Albert Einstein is history's most iconic scientist. His General Theory of Relativity plugged the gaps in Newton's work and described what gravity is – or, perhaps more accurately, it described what gravity is not: it refuted that it was a force at all. Einstein explained gravity by explaining it away.

Einstein was intrigued by the 'ghosts' that had appeared in Newton's gravitational machinery during the 19th century, as measurements became more precise. Astronomers who charted the positions of the planets watched Mercury and Uranus continually drift from the orbits predicted by Newtonian mathematics. Initially, the astronomers thought that undiscovered planets must be pulling the errant planets out of their orbits, and in the case of Uranus they were correct. On 23 September 1846, Neptune was discovered at almost exactly the position calculated for it by Urbain Le Verrier of the Paris Observatory. This was the first time a celestial object had been predicted to exist from a calculation. In a sense, it was the first time that astronomers postulated the existence of a piece of 'dark matter' –

thing unseen that betrays its presence by the gravitational effect it has on nearby celestial objects. The success of Neptune's discovery began a flurry of activity to find the planet that was now firmly thought to be pulling Mercury from its path. So certain were astronomers that a planet must exist between Mercury and the Sun they even named it: Vulcan.

However, there is no planet Vulcan. Instead, the movement of Mercury is due to an unexpected facet of gravity that Newton's theory does not take into account. Only when Einstein set about explaining the nature of gravity did he come across the remarkable reason for Mercury's motion.

The fabric of space

Einstein's great insight was the concept of the 'space–time continuum'. This is a fabric – for want of a better word – that stretches through all space, in all directions, and includes time as the fourth dimension. In Newton's theory of gravity, both space and time were imagined to be rigid frameworks, absolute and invariant; they were also quite separate concepts. In general relativity, space and time constitute a flexible continuum that can be stretched and warped by the presence of matter and energy. The warping of space–time can affect time as well as space, leading to a number of brain-bending consequences such as time dila-tion (see *Can We Travel Through Time and Space?*).

Einstein used the idea of the space–time continuum to explain gravity, stating that it was an effect created by the warping of space–time in the presence of matter. Think of a suspended sheet of rubber onto which is placed a heavy object. The object warps the rubber sheet, forming a curved depression. Any smaller object that is placed nearby will roll down the depression, spiral around the heavy object and eventually collide with it. This is analogous to the way in which gravity acts. In this scenario, the two-dimensional rubber sheet is the equivalent of the four-dimensional space–time continuum.

An often-quoted saying is that 'matter tells space–time how to curve, and space–time tells matter how to move'. But where is this curvature in our Universe? This is where the concept gets a little tricky – the curvature must be in a dimension of space that we are unable to perceive directly. We are familiar only with our three dimensions of space: up and down, left and right, in and out. Einstein included time as a fourth dimension and, further, asked that we accept that the gravity-generating curvature of space takes place through an additional dimension of space–time. Although we cannot see this other dimension explicitly, we feel it as if it were a force acting upon us, and we call it gravity. This is not as outlandish as it may at first sound. In fact it is very similar to centrifugal force, the rotationally generated effect that we first met in the context of planet formation (see *How Did the Earth Form?*).

Feeling the force

When we are in a car that takes a tight bend, centrifugal force makes it seem as if we are being pulled outwards. Although we feel as if a force is pulling us outwards, it is actually an illusion brought about because of inertia – our resistance to a change in direction. Our bodies would like to continue travelling forwards in a straight line but the car pulls us through a different dimension, and we perceive this change of direction as if it were a force acting on us. Imagine the car is remotely controlled and the windows are blacked out. Now, you have no outside reference against which to gauge your motion. Unbeknownst to you, the car is turned into a bend. Inside, you cannot see that you are turning but you feel the centrifugal force pulling you outwards. This is used to great effect in fairground ghost train rides when the cart you are riding suddenly veers off around an unexpected curve. So, too, according to general relativity we feel a 'force' of gravity because we are moving through the curvature of space in a dimension that we are unable to perceive directly.

A related concept that is central to general relativity is the 'principle of equivalence', which states that a gravitational field is indistinguishable from acceleration. Even in the 16th and 17th century philosophers were beginning to recognize this equivalence. Galileo Galilei proved that the speed at which an object falls does not depend on the amount of mass it contains. He did this by rolling balls of different masses down inclined slides and noticing that the balls always took the same length of time to hit the floor. He reasoned correctly that a feather takes longer to fall than a lead weight not because of the feather's smaller mass but because its structure is better supported by the air. Apollo astronaut Dave Scott performed a beautiful demonstration of this on the Moon in 1971 when he dropped a hammer and a feather at the same time. In the absence of air resistance, the feather fell at the same rate as the much more massive hammer, and both hit the lunar soil at the same time. This proved that not only objects of different mass but also objects of different composition are accelerated equally by a gravitational field. Einstein's extension of this in 1907 to the *equivalence* of gravity and acceleration is perfectly demonstrated with his famous 'thought experiment' involving a lift.

Einstein's lift

Imagine that you are in a lift, totally enclosed with no windows. When the lift is stationary, at whatever storey, gravity pulls you down as you stand on the floor. If someone were to cut the cables, the lift would fall and you would suddenly feel weightless. Any tiny body movement would result in you floating away from the floor because you would be falling at the same speed as the lift. This is how astronauts feel when they are in orbit. Inside the lift, you would continue to float until you hit the ground and your experiment came to a gruesome conclusion.

Now take the same lift into outer space, well away from any gravitating object. This time, you feel weightless because there are no gravitational forces acting on you and you are floating through space at the same speed as the lift. This is entirely equivalent to the lift falling on Earth. If you were trapped inside, you would be unable to distinguish between the two cases.

Finally, imagine strapping a rocket motor to the base of the lift in space and turning it on. The lift accelerates but your inertia does not want you to move, so you drop to the floor of the lift, which then pushes against you to accelerate you to the same speed as the lift. To you, this feels like the force of gravity experienced when you are standing in the lift on Earth; the faster the acceleration, the greater the force you feel – it is like experiencing a very great gravitational force. Indeed, in fast-moving jets and rockets, this force is often called the 'G-force'.

The principle of equivalence encapsulates the results of the lift 'thought experiment' by stating that acceleration is indistinguishable from a gravitational field. Once Einstein had accepted this, general relativity fell into place. His predictions are identical to those of Newtonian gravity when the gravitational force is weak or, in the language of relativity, when the curvature of space–time is shallow. However, when gravity becomes stronger and the curvature becomes more pronounced, general relativity predicts corrections to the way gravity acts on celestial objects. This is how Einstein explained Mercury's wayward orbit. Unlike the other planets, Mercury is close enough to the massive Sun for the curvature of space to be an important factor and so general relativity was needed to correctly calculate its movement. It was an early success for Einstein's theory, but not the final proof. As important as resolving a known problem is for a theory's credibility, the real test is whether it can predict something totally new. Einstein did not fail in this, either. He saw that, according to his theory, gravity should bend starlight much more than predicted by Newton's theory. The difficulty was, how he could prove it.

Gravitational lensing

To Newton, gravity could only affect objects with mass and, since light was a massless ray of energy, he considered it should pass through a gravitational field relatively unaffected. However, Einstein predicted that light would have to follow the contours of space–time, just as certainly as did planets and moons. His calculations showed that, when passing through a gravitational field, light would be deflected slightly from its path, in the same way a golf ball is deflected if it just clips the hole.

The concept is called 'gravitational lensing', and further calculations showed that the only object in the Solar System capable of bending starlight by a measurable amount is the Sun. The only time astronomers can see stars close to the Sun is during a total eclipse when the glare is blocked by the Moon. In 1919, Arthur Eddington led an observing party to the African island of Principe in the Atlantic Ocean to take the necessary measurements during the upcoming solar eclipse. Although cloudy in the morning, the weather cleared up in time for the dramatic total phase of the eclipse. In the sudden darkness, Eddington took his photographs, then measured the positions of the stars on the developed

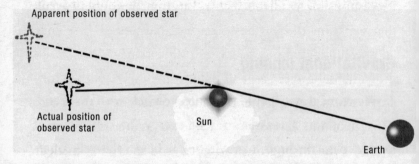

Gravitational lensing: starlight is deflected by the gravity of the sun

plates and compared them to photographs taken when the Sun was nowhere near the same stars. He found that the stars had apparently moved from their expected positions, just as Einstein had said they would if light were deflected by gravity.

The conclusion was inescapable: general relativity was correct. As the news spread around the world Einstein became a superstar, but this tremendous success has come at a price to modern physics: we cannot yet, almost a century later, integrate general relativity's explanation of gravity into our description of the other forces of nature. If we could, we might at last be able to establish the so-called 'theory of everything'.

Is gravity a force after all?

If we include gravity, there are four fundamental forces of nature. First there is the familiar force of electromagnetism; it is responsible for all the electrical and magnetic phenomena that we know of, and has been well-harnessed by science and technology since the late 19th century. Two other fundamental forces came to light in the early 20th century as physicists succeeded in probing the nucleus of the atom. These two nuclear forces are known, rather unimaginatively, as the 'strong nuclear force' and the 'weak nuclear force'. The strong nuclear force holds an atomic nucleus together and must be overcome in order to split an atom or to fuse two together. Hence, it is the strong nuclear force that allows energy to be liberated by stars and from atom bombs. The weak nuclear force governs certain forms of radioactive decay.

Taken as a force, gravity would be the weakest of all the fundamental forces. An easy way to demonstrate this is to watch an iron nail leap up towards a handheld magnet, proving that the magnetic force generated by the magnet in your hand is able to overwhelm the gravitational force created by the entire Earth. Nevertheless, it is the force of gravity that sculpts the Universe on its largest scales because it acts over vast distances, whereas the other forces are limited in the range over which they act. The two nuclear forces are confined to the width of an atomic nucleus; the electromagnetic force, although longer in range, tends to cancel out over large distances because it has both positive and negative charges, giving rise to repulsion (between like charges) as well as attraction (between opposite charges).

Physicists can explain the force fields associated with electromagnetism and the two nuclear forces as an exchange of tiny short-lived particles, called virtual particles. Gravity, on the other hand, can only be explained as a large-scale curvature of space. Many physicists suspect, and hope, for the symmetry and completeness of their theory, that gravity will eventually be found to be a real fourth force field, exchanging tiny virtual particles called gravitons. If that proves to be so, Einstein's curvature of space will turn out to be a scientific metaphor, useful for visualizing gravity before the true explanation was found.

The leading candidate for a theory that can unify gravity with the other forces of nature is known as 'string theory'. It replaces subatomic particles with minuscule bits of wiggling 'string' and uses these to conjure up all the particles of nature, including gravitons. String theory extends Einstein's ideas about other dimensions by having the strings wiggling through higher dimensions of space–time. We see the strings as point-like particles because we cannot perceive these other dimensions. Yet, for all the confidence in string theory, it is far from proven. The mathematics is so labyrinthine that even experts are having trouble relating the ideas to phenomena that might be observed with experiments and so enable the theory to be tested. Thus, finding a way to join gravity with the other forces of nature remains as difficult as ever; the difference in scale between minute virtual particles and large-scale curvatures of space–time is proving too great. Some new clue is needed to bridge the gap and, in the hope of providing this, physicists and astronomers are looking for any discrepancies between what general relativity predicts and what is actually observed. Such discrepancies would be rather like Mercury's deviation from Newton's predictions. They would be the signposts to a new understanding of gravity that can be dovetailed with our understanding of the other forces. The trouble is that no one can yet find any discrepancies.

Little green men

Einstein himself thought that his theory was likely to fail in extremely strong gravitational fields; in the 1950s, a class of celestial object was found that generated just such a strong gravitational field. That era saw the burgeoning use of radio telescopes, and a Cambridge graduate student, Jocelyn Bell, discovered a pulsating radio signal that was clearly celestial in origin because it appeared in exactly the same place in the sky night after night. It pulsed on and off, as regular as clockwork. She called the source LGM-1, which stood only half-jokingly for Little Green Men. Before long, however, the theoreticians showed that it was most likely to be the spinning super-dense remnant of an exploded star. Such a neutron star is even smaller than its cousin the white dwarf (see *What Are Stars Made From?*).

Whereas white dwarfs are about the size of the Earth and hold as much mass as the Sun, a neutron star is the size of a small asteroid – just 10 to 20 kilometres (6 to 12 miles) in diameter – and contains several times the mass of the Sun. The whole neutron star is packed with matter as tightly as in an atomic nucleus, hence its enormous density and very strong gravitational field. Radio astronomers tend to call them 'pulsars' because, as a neutron star spins, it can sweep powerful beams of radio emission through space, and, like a lighthouse beam, the radio signal appears to pulsate on and off as it sweeps over the Earth.

In 2003, a significant discovery was made: a double pulsar – two pulsars in orbit around each other. The pair is remarkable because the stars are separated by just 800,000 kilometres (500,000 miles), almost 90 times closer together than Mercury and the Sun, and speed around each other in just 2.4 hours. Einstein predicted that in such a strong gravitational field, orbiting objects would lose some of their energy and that this would be radiated away as a gravitational wave in the space–time continuum, rather like a ripple in a pond. By timing the contraction of the stars' orbits around each other, astronomers have calculated the amount of energy the double pulsar is losing. They have found that it is indeed the amount that Einstein's theory says it should be, at least down to the level that the telescopes can measure. The pulsars are moving closer to one another by seven millimetres every day. As the two get ever closer, so the rate of their energy loss will increase. Eventually, in about 85 million years, the two pulsars will collide, producing a cataclysmic explosion that will bathe much of the Galaxy in gamma rays. The accuracy with which general relativity predicted this energy loss was again great news for Einstein's theory, but not such good news for physicists hoping for a clue to point them towards fresh ideas about the nature of gravity and how to integrate it with the other forces of nature.

The ultimate test

Most modern tests of general relativity concentrate on testing the principle of equivalence, hoping to find a difference in the way that gravity acts compared with non-gravitational accelerations. Only in general relativity is the principle of equivalence exact. In string theory and in other attempts to unify gravity with the other forces the equivalence is only approximately true, which means that with sufficiently sensitive measurements deviations should be found.

One of the most promising experiments has been running for more than 40 years, made possible by the Apollo Moon landing missions. It is known as 'Lunar Laser Ranging'. The Moon provides a remarkable gravitational laboratory – a giant test mass close enough to be within reach. During their observing runs, astronomers at the Apache Point Observatory, New Mexico, fire a powerful laser beam at the Moon. They target suitcase-sized reflectors, left on the lunar surface by three of the Apollo missions and two Russian missions. It is a job requiring tenacity and patience because out of every 300 million billion photons of light that the astronomers send to the Moon, just five find their way back to the observatory's waiting telescope. The rest are lost to the atmosphere of the Earth, or miss the lunar reflectors entirely to be absorbed in the lunar soil or bounced off into a random direction in space. From the tiny numbers returning, the astronomers have been able to measure the movement of the Moon to an accuracy of a centimetre or two, and this has allowed them to calculate that the Moon is moving as Einstein said it should, to within one part in ten trillion. So general relativity *still* holds.

Recent upgrades in the ground-based equipment have allowed the movement of the Moon to be measured to an accuracy of millimetres. This will place general relativity under even more stringent tests, and there are many physicists anxiously waiting for the results. But for now, Einstein looks remarkably right; one might even say disappointingly right, because the requirements of general relativity prevent progress to what many feel would be a deeper understanding of the cosmos.

What is a black hole?
Gobbling monsters, evaporating pin pricks and balls of string

'Black holes' conjure up curiosity and confusion in equal amounts. The concept sprang out of the mathematics of Einstein's general relativity theory but has only recently attracted huge popular attention. Often black holes are portrayed as all-powerful destroyers that capture and crush everything around them. Thankfully for the Universe at large, this is not quite true.

Black holes could be said to be the weirdest things known to exist. The theoretical possibility of their existence has been around for almost a century, and we now have 30 years of strong observational evidence, yet astronomers still do not fully understand them.

A black hole containing four or five times the mass of the Sun would occupy a spherical volume just a few kilometres across. This would curve the fabric of space so sharply (see *Was Einstein Right?*) that the strength of gravity would change greatly even across the length of a human body. If you were in the vicinity, your feet would accelerate much faster than your head, stretching you on the gravitational equivalent of a medieval rack. The phenomenon is an extreme version of a tidal force (see *Why Do the Planets Stay in Orbit?*) and would eventually pull you apart in a process known, with black humour, as 'spaghettification'. As your constituent atoms passed through the 'event horizon', the invisible boundary after which there is no escape, you would become part of the riddle that so far defies solution: what is inside a black hole?

Dark stars

The first inklings of the possibility of black holes – though they were not named as such – came as far back as the 18th century when geologist John Michell wrote to the Royal Society in London suggesting that a star 500 times larger than the Sun would generate so much gravity that not even light could escape from its clutches. Michell was inspired by the natural philosophers of his time, who were succeeding in measuring the finite, constant speed of light.

The third Astronomer Royal, James Bradley, was the first to calculate the currently accepted value for the speed of light. Working at Greenwich, London, in 1728 he detected a strange movement in the position of the stars, amounting to just 1/200 of a degree. At first he thought that he had detected stellar parallax (see *How Big is the Universe?*) but soon realized that all the stars he measured had the same angular displacement, and so the effect could not be parallax, which would have varied with the star's distance from Earth. He proposed that the movement was caused by the finite speed of light. In the same way that you have to slightly tilt an umbrella in front of you when walking through a shower because your motion makes the raindrops appear to be approaching at a slightly diagonal angle, so he was having to angle his telescope to compensate for the Earth's motion through space. The angle of the tilt allowed Bradley to calculate the speed of light in relation to the speed of the Earth. He computed a figure of 186,000 miles per second (or around 300,000 kilometres per second).

Michell took Bradley's figure and used Newtonian gravity to estimate the size of body needed to have an escape velocity (see *Why Do the Planets Stay in Orbit?*) equal to the speed of light, and came to his estimate of 500 times the mass of the Sun. The idea sparked a debate that rumbled for a number of years as astronomers mulled the possibility of such 'dark stars'. Eventually the natural philosophers decided that Newton's laws precluded light from being affected by a gravitational field and so light would always leave a celestial object, no matter how strong its gravity.

The matter rested for a couple of centuries until Einstein published his General Theory of Relativity in 1915 and showed that gravity did indeed affect light (see *Was Einstein Right?*). Less than two months after Einstein's publication, German mathematician Karl Schwarzschild found that Einstein's equations allowed celestial objects to become so dense that they create gravitational traps. The size of each trap, known as its 'Schwarzschild radius', is determined by the mass inside. For example, a black hole containing the mass of the Earth would have a Schwarzschild radius the size of a small coin, whereas a black hole containing billions of times more mass than the Sun would be as large as our Solar System. Once any object, or even light, had passed beyond this Schwarzschild radius – or 'event horizon' – it could never escape.

Astronomers were forced to accept that black holes *could* exist, but the dilemma was how they could possibly observe something that emitted no light or any other type of radiation. The solution did not come until the early 1970s, when the first X-ray telescopes were lofted into space and revealed an extraordinarily bright X-ray source in the constellation of Cygnus, some 8000 light years away. After much analysis, it was decided that the source of X-rays was a superheated cloud of gas spiralling into a black hole, dubbed Cygnus X-1. As the gas accelerated in the immensely strong gravitational field outside the event horizon, it was heated to millions of degrees and began to emit X-rays.

Since then numerous observations have been made of a black hole 'feeding', for example by ripping apart a companion star. The blue supergiant star HDE 226868 is about 30 times more massive than the Sun and 400,000 times brighter. The black hole next to it contains only between 5 to 10 times the mass of the Sun but has a gravitational field so strong that it has pulled the blue supergiant into an egg shape and is now greedily stripping it of gas. The gas falls from the giant star and enters a brief spiral orbit around the black hole, forming what astronomers call an accretion disc. In the maelstrom, magnetic fields compress some of the gas into jets that seem to shoot away from oblivion. Most of the gas, however, ends up heading into the black hole like water spiralling down a plug hole and then disappears forever.

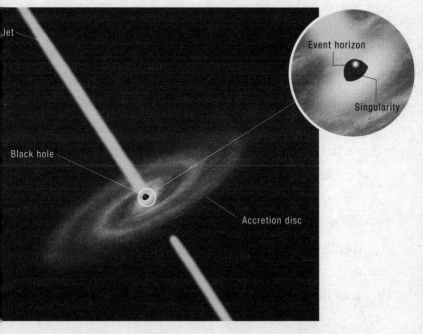

Anatomy of a black hole: a black hole is not a simple object but is made up of several components

Small, intermediate and supermassive black holes

Black holes such as Cygnus X-1 are known as 'stellar' black holes. They contain several times the mass of the Sun and are formed when very massive stars explode as supernovae (see *What Are Stars Made From?*). The supernova is triggered when the inert iron core of the star collapses to become a neutron star. As the outer layers of the star come crashing down, igniting the supernova explosion, the neutron star at the core is pummelled and some of the outer material is absorbed, increasing the core's mass, which can increase the gravitational field so much that the core becomes a black hole.

Whilst such stellar black holes are the most prevalent kind, there are now known to be other, larger, black holes with different formation mechanisms. The next size up is termed an 'intermediate mass' black hole. In common with the stellar black holes these orbit the centre of their galaxy and contain a few hundred or a few thousand solar masses. Astronomers are not sure how these form; possibly they result from several stellar-sized black holes merging together.

Thirdly, there are the 'supermassive' black holes, containing anything from millions to billions of times the mass of the Sun. A supermassive black hole is thought to sit at the centre of *every* galaxy, but taking up no more volume than an average solar system. In 90 percent of galaxies the central supermassive black hole is inactive but, in the other ten percent, it is constantly feeding from surrounding celestial objects and this drives an extraordinary engine of activity that can be seen across billions of light years of space.

Active galaxies

Vast quantities of radiation are released by an active galaxy, all derived from matter heating up before it plunges into the supermassive black hole at the galaxy's centre. The most powerful active galaxies generate more energy per second than a trillion Suns, with the result that the active nucleus outshines the rest of the galaxy by a hundred times or more. This brilliance masked the nature of an active galaxy for some time; when astronomers caught their first glimpses of them during the 1950s, they saw the star-like active cores and assumed that they were peculiar stars in our own Galaxy. They called them 'quasi-stellar' objects, from which the present name quasar is derived.

The true identity of quasars was revealed in 1962, when astronomers discovered that they are in fact incredibly distant, so could not be stars but had to be intensely powerful galaxies. As the distribution of these highly active galaxies was charted, it was revealed that *all* quasars are in the far reaches of the Universe and none exist nearby. Because light takes so long to travel across space this means that quasars are ancient objects – they seem to have populated the Universe in their greatest abundance around 10 billion years ago. This leads astronomers to conclude that quasars are a phase that every galaxy passes through, brought about when the supermassive black hole at its centre has a lot of matter to consume.

Less powerful active galaxies can be found throughout the Universe at all distances. Some may be ageing quasars whose food source is almost used up. When the supermassive black hole finally devours everything within its reach, the active galaxy quietens to become a normal galaxy, such as our own. But there is nothing to stop the black hole coming back to life if more matter falls into its clutches. According to calculations, one medium-sized star like the Sun wandering too close to the galactic centre is all that would be needed to re-ignite the activity and keep the black hole spewing energy for a year. So, the current population of active galaxies must be transient. If we came back in a million years' time, some presently active galaxies would have become inactive, whilst other currently quiet ones would be blazing with energy.

Black hole silhouettes

Seeing a black hole that is *not* feeding was once thought to be impossible but that view is now changing, thanks to technological advances in radio telescopes. Within a decade astronomers anticipate being able to look for a black hole's 'silhouette' against the background of bright stars. This is far from easy. Our own Galaxy's central supermassive black hole, known as 'Sagittarius A*' (pronounced A-star), is estimated to be about 4.5 million solar masses, all squeezed into an event horizon some 27 million kilometres (17 million miles) in diameter; this is only about half the distance of Mercury from the Sun. From our vantage point on Earth, the silhouette of Sagittarius A* would appear no larger than a football on the surface of the Moon.

By combining the simultaneous observations of radio telescopes across the world, however, astronomers hope to be able to distinguish the silhouette of the black hole, and even to catch sight of a few clouds of gas straying into its gravitational maw. By watching the paths that these clouds take to their destruction, it will be possible to measure how fast Sagittarius A* is spinning. If, as is theorized, a black hole is spinning, general relativity tells us that it will create a vortex in the fabric of space – imagine the twisted mass that can be created by spinning a spoon in a jar of honey. This twisted region around a black hole is called the 'ergosphere'. Observations of gas clouds caught in the ergosphere of Sagittarius A* would open up a whole new way to investigate not only the black hole, but also the validity of general relativity in such an extreme environment.

Evaporating black holes

There may be a fourth type of black hole, at the opposite end of the scale. These are tiny 'primordial' black holes, thought to have been created during the Big Bang when the space-time continuum was so crushed that minuscule regions could seal themselves off from the rest of the cosmos. Such a primordial black hole has never been detected although, in the 1970s, the celebrated physicist Stephen Hawking suggested a way in which we might see them, and at the same time prove that black holes are not completely black. He did so using another cornerstone of modern physics: quantum theory.

First propounded in the early decades of the 20th century, quantum theory describes the Universe on its smallest scales. Central to its explanation is that energy comes in discrete packets, so what may look like a beam of light is actually made up of a multitude of tiny particles called photons. Each one of these photons carries a well-defined amount of energy; for example, a photon of blue light carries twice the energy of a photon of red light. Quantum theory also tells us how particles behave, and one of its tenets is that a particle is difficult to pin down to a specific place. Physicists can calculate where they expect a particle to be, but it could be just as easily somewhere else in a small region around this calculated position. This means that particles travelling close to a boundary can sometimes appear to have spontaneously jumped across it, a phenomenon known as 'tunnelling'. It is manifest in actual observations: it makes fusion in stars possible at temperatures lower than would normally be needed, because the closely packed atomic nuclei occasionally find themselves close enough to their neighbour to fuse together and release energy.

According to Hawking, a particle can tunnel from the interior of a rotating black hole and escape, thus lowering the black hole's mass. As the black hole loses mass, so the process continues and speeds up, until the black hole disappears in a sudden release of gamma rays. The most likely black holes that 'evaporate' in this way are the primordial ones, because they will be evaporating faster than they are capable of consuming. Yet, although gamma ray satellites have been on the lookout for this behaviour, nothing has yet been seen. It is possible that the Large Hadron Collider in Switzerland will create minute black holes in particle collision processes, which may evaporate in a fraction of a second and give scientists their first glimpse of this process.

The problem of the singularity

Although black holes are now an accepted part of the pantheon of celestial objects, there is still unease about them in astronomical circles. One of the uncanny things is that mathematically they share a striking similarity to the Big Bang. At first, there may not seem to be anything in common between a black hole, which crushes things out of existence, and the Big Bang, which created the Universe and set it expanding. However, to a mathematician they share an identical feature: a point of infinite density and zero volume, known as a 'singularity'. Inside a black hole, the singularity is presumed to be the last resting place of matter because gravity crushes that matter into smaller and smaller volumes. This presents a problem for physicists because as the volume approaches zero no theory can be used to study the resulting singularity.

There is also something known as the black hole 'information loss' problem. To the outside Universe, only three properties of the black hole are visible: its mass, its electric charge and its angular momentum (rotation). Any other information, such as what fell in, appears to be lost. This goes against the grain of one of the deepest principles of physics: that of reversibility. If you drop something into a bowl of water and let it dissolve, in principle someone could analyse the water, and identify what has been dropped in there. They could even separate the substances again by boiling off the water and thus reconstruct the original material. This is reversibility. In the case of a black hole, once something crosses the event horizon we cannot then discover what it was, let alone recover it. All the stars and planets that have been devoured have been erased from the Universe: their composition, their temperature, their density – all are gone. Not even particles tunnelling out, according to Hawking, can provide us with the missing information.

Physicists hope that their efforts to unify gravity with the other forces of nature using string theory (see *Was Einstein Right?*) will allow them to investigate this conundrum. Indeed, string theory may even have given them their first clue.

Black hole or fuzz ball?

String theory suggests that black holes do not have singularities but that their volume from the centre out to the event horizon is a highly compressed ball of subatomic 'strings', the fundamental building blocks of nature that according to the theory give us particles of matter. These compressed strings would store the fundamental information about the objects that have fallen into the black hole, so no information has actually been lost.

In this view, matter does not pass *through* the event horizon on its way to the singularity; instead it compresses itself onto the surface of the 'fuzz ball' and merges with the other strings. Think of it as layers of paint, but instead of each successive layer overwriting the last one, they run together, and the black hole's Schwarzschild radius grows a little larger to make room for the latest arrivals. In relativity this is explained as the curvature of space becoming a little steeper, because the black hole has swallowed more mass. In string theory, the black hole simply expands a little to accommodate the new information.

Black holes are the most extreme celestial objects that we know and, as such, continue to challenge our understanding of the most fundamental laws of physics. Even if a black hole is not truly black, and not a hole but a fuzzy ball of quantum strings, everyone agrees about one thing: you definitely would not want to fall into one.

How did the Universe form?
Picturing the Big Bang

The importance of the Big Bang cannot be overstated: it is thought to be the 'moment' at which space and time themselves began. Physicists can say nothing about what created the Universe, or why, but our natural curiosity leads us to try to picture how it all began.

It seems barely credible, but the definitive proof of the Universe's moment of creation was at first dismissed as pigeon droppings. In 1964, two researchers from Bell Laboratories, New Jersey, began to pick up a strange signal through a radio telescope. At first, they thought it might be interference from New York City. When they investigated and rejected that idea, they noticed that the telescope structure was home to a pair of nesting pigeons who had peppered the device's sensitive surfaces with droppings. So they cleaned the telescope and shipped the birds across the state. Pigeons being pigeons, however, they flew home and took up residence once more in the telescope. It was necessary to resort to a final solution: a man with a shotgun. Yet even with the pigeons despatched and the telescope scrubbed with the pigeons despatched and the telescope scrubbed clean again, the mysterious signal remained. The researchers, Arno Penzias and Robert Wilson, began to think that it might be coming from space.

Unknown to them, a group of astronomy theoreticians had predicted its existence almost two decades earlier. In 1948, the Ukrainian physicist George Gamow had been exploring the consequences of Georges Lemaître's 1927 idea that the entire Universe exploded from a single compacted 'atom' some time in the remote past (see *How Old is the Universe?*). Gamow used this idea of a Big Bang to show that it explained the overwhelming quantity of hydrogen and helium in the Universe. He calculated that, if the Universe began with nothing but the simplest chemical element, hydrogen, then the intense heat of the Big Bang would have fused a quarter of it into helium: almost the exact proportions that astronomers were seeing in their inventory of the Universe. Gamow also made the prediction that radiation from the fireball that forged the helium would still be lingering in the Universe today as an all-pervasive blanket of microwaves. So, when Penzias and Wilson began talking about detecting microwave static from outer space, they caused a sensation. Cosmologists quickly pointed out that the pair had discovered nothing less than the residual radiation from the Big Bang.

The beginning of everything

Current physics cannot describe the very beginning of the Big Bang, because it cannot deal with the tiny fractions of time and space that would need to be considered. The smallest unit of time that physics can presently handle is 10^{-43} seconds: a decimal point followed by 42 zeros and a one. It is known as the 'Planck time', after Max Planck, the father of quantum theory (see *What is a Black Hole?*). All we can say is that during this time, termed the 'Planck era', everything that we see in the Universe today was squeezed into a tiny dot, smaller than an atomic nucleus. The four fundamental forces of nature – gravity, electromagnetism and the strong and weak nuclear forces – were indistinguishable from one another, and the 'dot' was already expanding. To fully picture the Planck era, a theory of quantum gravity is needed, such as string theory (see *Was Einstein Right?*).

As the Planck era ended, so gravity became a distinct force and physics as it is presently understood took over. However, the temperature and pressure were so extreme that matter and energy were entirely interchangeable; particles would form spontaneously from the writhing energy. Every time this happened, out too would leap particles of unusual stuff: antimatter.

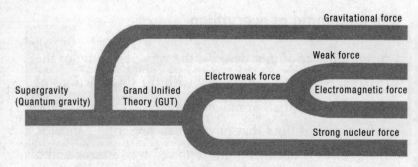

The unification of natural forces: the four fundamental forces of nature were once joined together

Antimatter

The concept of antimatter first found its way into the minds of physicists in 1928 when British physicist Paul Dirac discovered an equation that correctly described the behaviour of electrons moving at high speed, but predicted that 'mirror-image' electrons should exist as well. These other electrons would be identical in mass but would carry positive electrical charge instead of the normal negative charge. Just four years later came the experimental proof, with the discovery of a positively charged electron (later called a positron) in a shower of particles coming from space. Dirac extended the idea to all particles of matter, and coined the umbrella term of 'antimatter' for the oppositely charged counterparts. One of antimatter's properties is that, should it run into its mirror-image piece of matter, both will transform into pure energy. For example, the collision of a positron with an electron will annihilate both, giving out a pair of gamma rays.

This leads to one of the most vexing questions for modern cosmology: why is there any matter left in the Universe? Since an antimatter counterpart is predicted to accompany every particle's creation, it should mean that everything eventually annihilates back into energy. Yet the existence of stars and planets and galaxies shows clearly that there is a residue of matter.

The solution is bizarre: mathematical calculations reveal that for every billion particles of ordinary matter created after the Big Bang, there were only 999,999,999 particles of antimatter formed. These annihilated with the corresponding matter, leaving a single orphaned particle of matter. This process repeated over and over again in the early Universe, building up enough matter particle by particle to make all the celestial objects. It means that for every particle of matter that exists today, there were once a billion other particles, but all of these were annihilated back into energy – which eventually became the microwave background radiation we see today.

Cosmic inflation

In the period following the Planck era when matter was being created, cosmologists believe that the Universe underwent a sudden intense expansion that they call 'inflation', which drove the Universe to balloon by a factor of 10^{50} in a time of just 10^{-32} seconds. The observations that led astronomers into this thinking are encapsulated in two thorny cosmological problems.

The first is the horizon problem. This is best explained by reference to the cosmic microwave background, which shows that the temperature on one side of the Universe is the same as on the other. The energy carried by the microwave radiation governs the temperature of space by heating any molecule or atom that gets in its way to the same temperature of about 2.7 kelvin (-270.3 degrees Celsius). The conundrum is why it should be the same everywhere, when the two far sides of the Universe cannot yet know of the other's existence. In scientific parlance, they are said to lie outside each other's observable horizon. Separated by at least 26 billion light years and probably much more (see *How Big is the Universe?*), the opposite sides of the Universe have not had time, in the Universe's lifetime of 13.7 billion years, to exchange energy and so equalize their temperature. Yet, the entire Universe displays the same temperature, no matter where astronomers look. They do not know how this could happen, so they call it the horizon problem. The inflation theory supplies a possible answer because it states that the entire volume of our Universe came from a vanishingly small region that suddenly grew in size and spread the same temperature across space.

Second is the flatness problem. Einstein's proposition that matter curves the fabric of space means that the Universe should have an overall curvature, determined by the total amount of matter and energy it contains (see *What Will Be the Fate of the Universe?*). But as far as anyone can tell, the Universe is completely 'flat' on the largest scales, an extremely unlikely outcome of Einstein's equations that represents a perfectly balanced cosmos. Again, inflation could provide a solution because whatever intrinsic curvature the Universe has, it has become spread over such a large scale that we can no longer perceive it. This is exactly like the way the ground beneath our feet appears flat even though we know it is part of the curved surface of the Earth.

Whilst inflation helps with these problems, the query remains as to why the Universe would have inflated in this way. It may be linked to the way the strong nuclear force 'broke away' from the electroweak force (the still-united weak nuclear force and electromagnetic force). This separation may have produced energy and driven the inflation, but this is uncertain and the subject of much debate between physicists.

Post-inflation

By the time the Universe in cosmological models reaches a millionth of a second in age, there is more confidence about what is going on. This is because powerful particle accelerators, such as the Large Hadron Collider in Switzerland, can re-create the high-temperature, high-pressure conditions by smashing particles together with great energies and analysing the debris.

At a microsecond, the Universe was filled with subatomic particles called quarks, together with antiquarks and gamma-ray photons. Quarks are the smallest building blocks of ordinary matter. These began to gather themselves together to form protons and neutrons, which in turn went on to form the atomic nuclei of today. As time ticked forward, the Universe continued to expand – although much less rapidly than in the inflation era – lowering the density of matter and energy, and so reducing the temperature and pressure. At around two seconds after the Big Bang, the electroweak force separated into the weak nuclear force and electromagnetism; now all four of the fundamental forces displayed their unique characteristics. This led on to the next great era of the Big Bang, which began about three minutes after creation, and ended perhaps a quarter of an hour later. It is known as the era of 'nucleosynthesis' and is the period George Gamow investigated mathematically during the 1940s.

For these few minutes, the entire Universe was somewhat similar to the interior of a star; in this hot, dense maelstrom, some of the protons and neutrons combined to form helium and lithium nuclei. When formulating his theory, Gamow originally thought that all the elements could be forged in the aftermath of the Big Bang, but as he looked at the calculations he realized that they only worked well for hydrogen, helium and lithium; subsequent investigations confirmed that the Big Bang would not build anything heavier than lithium. The existence of heavier atoms in the Universe today became a mystery that was solved in 1954, when British physicist Fred Hoyle showed that the nuclear furnace at the centre of a star is the only place where the rest of the chemical elements can be built (see *Are We Made from Stardust?*). As a result, we now know that heavy elements did not appear in the Universe until nearly a billion years after the Big Bang and this has implications for the nature of the first celestial objects that could form (see *What Were the First Celestial Objects?*).

The Universe then entered a relatively calm phase, becoming a sea of jostling particles and photons of energy. The photons, created by the matter–antimatter annihilations, continually collided with electrons, preventing them from bonding to the atomic nuclei, and thus creating a state of matter called plasma. Plasmas exist today inside stars and in clouds of gas surrounding high-mass stars; in the early Universe the plasma stretched across all space.

By the time the Universe was a year old, collisions between photons and particles had become less frequent because the continuing expansion gave the particles more space in which to move about. This allowed gravity to begin slowly pulling the particles into clumps. As time passed and the density of the Universe continued to decrease, around 380,000 years after the Big Bang one of the greatest watersheds was reached – the 'decoupling' of matter and energy. All of a sudden, the drop in number of electron-photon collisions meant that electrons could be captured by the atomic nuclei. Neutral atoms of ordinary matter formed, and space was cleared of its fog of particles. Some say that this was the point at which the Universe became transparent, because most photons could now travel all the way across the Universe without running into an absorbing particle. At this stage the photons made up a powerful sleet of X-rays; they have been continually redshifted by the expanding Universe ever since and now exist as the microwave background detected by Penzias and Wilson. In the future they will become weaker still, stretching out to become mere radio waves, the weakest form of electromagnetic radiation.

The Universe's baby pictures

Maps of the microwave background radiation in all directions have been analysed and found to contain minuscule variations in its temperature from place to place. These 'anisotropies' amount to differences of no more than a hundred-thousandth of a degree, but are highly significant. They are the imprints left on the microwaves by the clumps of matter that formed in the early Universe, around its 380,000th birthday, and show us the seeds from which galaxies and the large-scale structure of the Universe emerged.

To get a snapshot of the Universe at a younger age, astronomers must turn to tiny particles known as 'neutrinos', which appear during interactions involving the weak nuclear force. About two seconds after the Big Bang, the moment that the weak force separated from the electro-magnetic force, a gigantic burst of neutrinos would have been released. This should be all around us even now, just as the microwave background fills all space. Neutrinos, however, are far more difficult to observe than microwaves. They are ghostly particles that find it easy to slip right through detectors, in fact they stream unnoticed through the entire Earth. In the time it has taken you to read this page, trillions of them will have passed through you, into the Earth and out of the other side to continue their journey across the Universe.

The possibility of such flighty particles first presented itself in calculations during the 1930s, but it took almost three decades before a neutrino was captured in a purpose-built detector. The early versions of these detectors resembled giant swimming pools, but buried deep underground so that the surrounding rocks could shield them from unwanted particles. They were filled with water or some other detecting fluid and lined with sensors that would record the flash of light produced when a neutrino struck a molecule in the tank. Typically they would capture one or two neutrinos a month. Modern neutrino detectors can be found buried in the Antarctic ice, or sunk below the surface of the Mediterranean Sea. They still look for flashes of light produced by neutrino collisions, but they scan the ice itself, or seawater, for the faint flash that betrays each neutrino's passage. It is hoped that within a decade these detectors will provide a neutrino map of the entire sky. Unfortunately for cosmologists, however, it is not the low-energy neutrinos given out two seconds after the Big Bang that will be mapped, but high-energy ones generated from the explosion of stars. Nevertheless, this will be a step in the direction towards grabbing a picture of the Universe as it looked when it was two seconds old.

Cosmologists' ambitions do not stop there. The ultimate picture of the Big Bang may be possible if physicists can unify the forces and deduce the nature of the particles suspected to carry quantum gravity, the gravitons. By analogy with the weak nuclear force and neutrinos, the separation of gravity at the end of the Planck era would have generated a background of gravitons. There is speculation regarding a future graviton telescope, which would be able to detect these and so give a picture of the Universe just 10^{-43} seconds after the Big Bang. To all intents and purposes, it would be a picture of the Big Bang.

What were the first celestial objects?

The beginnings of the Universe as we know it

Astronomers call it the dark ages. It began around 380,000 years after the Big Bang, when atoms had just formed and X-ray radiation permeated space. There were no galaxies, no stars, no planets at that time – nothing that could shine light into the cosmos. Very gradually, gravitational attraction drew together clouds of matter and eventually the first celestial objects were born, some time between 200 million and one billion years after the Big Bang.

Whatever the first celestial objects were, they pumped so much light energy into space that they ionized almost every atom in the Universe by blasting away the electrons from their atomic nuclei. This created giant clouds of glowing gas that further lit up space. Thanks to the time it takes light to cross the vast tracts of space, those first luminous sources, more than 13 billion light years away, should still be visible with a sufficiently large telescope as tiny pinpricks of light.

Deep fields

The first attempt to see into the furthest reaches of the Universe was in 1995, when the orbiting Hubble Space Telescope pioneered the technique of 'deep field' astronomy. A ten-day observation was conducted that consisted of looking at a single patch of sky no larger than a tennis ball placed 100 metres away. The area chosen was one near the constellation of Ursa Major (also known as The Plough or Big Dipper) that appeared to be completely empty – nothing had ever been found there. After ten days of collecting light, however, the Hubble Telescope produced an image revealing 3000 celestial objects. The vast majority of them were small galaxies, at distances greater than 10 billion light years.

The image became known as the 'Hubble Deep Field' and gave astronomers their first real look at such extremely distant realms. Previously, observations with ground-based telescopes had only detected galaxies within 7 billion light years, about halfway across the Universe. These galaxies seemed indistinguishable from present-day galaxies, suggesting to astronomers that, however galaxies formed, they did it relatively quickly, building themselves into their mature shapes within the first six billion years of the Universe's history. Today's galaxies are classified according to their shape (see *What is the Universe?*). Elliptical galaxies are elongated balls of stars; spiral galaxies are flat with a central hub of stars and arms that spiral around the centre; barred-spiral galaxies each have an elongated central hub connecting to the spiral arms. There are also irregularly shaped galaxies.

The large numbers of small, distant galaxies in the Hubble Deep Field confirmed that today's large galaxies began as much smaller collections of a few million or fewer stars. They were either irregularly shaped or elliptical, and built themselves into larger galaxies by colliding and merging with their neighbours. As they grew in size, they eventually accumulated enough mass to develop appreciable gravitational fields with which to pull in gas from intergalactic space. As the gas plummeted towards the galaxy, it fell into orbit in a disc around the central hub of stars. When the disc accumulated enough gas, star formation spontaneously began within it and surrounded the galaxy with sweeping arms of new stars.

There are two subtly different patterns of spiral arms that can form. Each betrays the behaviour of the gaseous disc in that particular galaxy. First there are the 'grand design' spirals, consisting of two dominant spiral arms that can be traced outwards from the centre of the galaxy. Grand design spirals are caused by a rippling wave of matter that rotates around the centre of the galaxy. These ripples, or 'density waves', compress dust and gas as they pass, triggering star formation. By contrast, the second type of spiral galaxy, the

Types of spiral galaxies: grand design spiral galaxies (*left*) have well defined spiral arms whereas flocculent spirals (*right*) do not.

'flocculent' spiral, has a messy whorl of stars, which form without the intervention of density waves. As a cluster of bright stars forms, those stars closest to the centre of the galaxy complete their small orbits in a short time, while those further out take longer. This stretches the star-forming regions into truncated spirals; thousands of these contribute to the 'woolly' appearance of a flocculent spiral galaxy.

Mergers and acquisitions

If a spiral galaxy is left on its own, it will continually accumulate gas from its surroundings. Perhaps it will occasionally cannibalize a much smaller galaxy, but neither process will affect its overall spiral shape. However, should it veer too close to a similarly sized galaxy, the resultant collision will destroy the delicate spiral shape. As they merge, the two galaxies will lose all structure and the result will be a large fuzzy cloud of stars: an elliptical galaxy. This mighty collision will also force all remaining gas in the merged galaxies to transform into new stars, triggering a sudden explosion of star formation known as a 'starburst'. In a few hundred million years, a multitude of brilliant star clusters will be created until all of the gas is exhausted.

At the centre of the merging galaxies other dramatic events will be afoot. Both galaxies will contain a central supermassive black hole (see *What is a Black Hole?*) and these will both sink towards the centre of the merged galaxy, where they will draw each other into spiralling orbits. Their huge gravitational fields will interact, swallowing stars and throwing others into eccentric orbits. Eventually the black holes will meet and plunge together, releasing a torrent of energy that sweeps through the galaxy as a burst of radiation. The newly enlarged black hole will continue to consume clouds of gas, stars or planets that haplessly stray into its gravitational reach. For the first few million years following a merger, this can be a massive amount of material, and the merging galaxies will most likely become a quasar: a highly active, tremendously luminous galaxy. These once populated the Universe in great numbers but have dwindled to extinction, no doubt because the black holes that power them have devoured everything within their gravitational grasp.

Once the quasar eventually dies down, it becomes an ordinary galaxy with a dormant central black hole. Cosmologists believe that the Universe built its current quota of galaxies in this way. But the nature of the first step of the sequence – the origin of the collections of a few million stars – remains elusive.

Ultra deep fields

When the Hubble Space Telescope was upgraded with a new camera, astronomers tried another deep field observation. The 'Hubble Ultra Deep Field', as the new shot was called, covers an area of about one-tenth that of the full Moon and revealed 10,000 small galaxies. Most of them appear as they looked around 800 million years after the Big Bang, which is staggering, but still there was no sign of the very first, individual celestial objects. They must be more distant still, and too faint to be seen by the Hubble Telescope.

So astronomers have had to turn to the theorists, who use computers to model what were likely to have been the first celestial objects drawn together by gravity. There seem to be two possibilities: either they were stars, gigantic by our modern standards; or they were black holes, already busily sucking in gas that would radiate furiously as it fell into oblivion. Whichever they were, they were the objects that clustered together to become the galaxy building blocks, and prepared the Universe for the formation of other celestial objects. As stars and black holes would do these jobs in different ways, it is crucial for cosmologists to determine which it was, in order to understand the subsequent development of the Universe.

Megastars or black holes?

Of all the celestial objects, stars are the ones that exert the biggest influence over the Universe's chemical composition (see *What Are Stars Made From?*), and none have affected it more than the earliest stars. During the 'dark ages' before the first luminous objects, all that existed was a diffuse sea of atoms: roughly three-quarters of it hydrogen, one quarter helium, with a seasoning of lithium. No other chemicals yet existed, and computer models suggest that this lack of variety had a tremendous effect on the first generation of stars.

As gravity pulls gas together to form a star, so the gas naturally heats up as its atoms are confined. This heat resists further compression and must be radiated into space in order for the star to continue pulling itself together. Calculations show that the heavy chemical elements are highly efficient radiators, whereas the light gaseous elements find it difficult to dissipate their energy. So in star formation today, the presence of elements heavier than lithium speeds up the collapse, allowing stars to form from relatively compact pockets of gas. This results in most stars containing less mass than the Sun. Back in the dark ages, however, the forming stars did not have the help of the heavy elements in losing heat, and this meant much more gas had to build up in order for gravity to become the overwhelming force. The first stars were therefore much larger than those found in the present Universe, with masses from several hundred to a thousand times that of the Sun. One of these megastars would be big enough to engulf all the planets in the Solar System, were it placed at the Sun's location.

It was first thought that these early megastars were the chemical factories of the Universe, with their vast nuclear furnaces transforming a fraction of their hydrogen and helium into the other chemical elements and then scattering it into space, where it could be incorporated into the next generation of smaller stars. But there is a flaw in this theory. Hydrogen is transformed by nuclear fusion in one of two ways: either in a series of collisions called the 'proton-proton chain' or through a reaction sequence known as the 'carbon-nitrogen-oxygen cycle' (CNO cycle for short). The proton-proton chain is the less efficient of the two but, because in any early megastar there was no carbon to begin the CNO cycle, it would have been forced to rely on the proton-proton chain. The difficulty is that the proton-proton chain would not have been able to generate enough energy to counteract the gravity of a star of that size, and so the star should have simply collapsed – and become a black hole. A black hole made in this way would have possessed from several hundred to a thousand solar masses and could certainly have been the seed around which a galaxy began to form. In effect, these early black holes would have been quasars in the making. Yet this picture cannot be completely right either, because light collected from a handful of extremely distant quasars that were shining just 900 million years after the Big Bang shows the telltale pattern of absorption lines (see *What Are Stars Made From?*) that betrays the presence of iron – an element that can only be formed in the heart of a massive star. So, at least some stars must have formed before these quasars.

It seems likely then that a mixture of stars and black holes constituted the first celestial objects. The only way to understand exactly what was going on back in those distant times is to find a way of seeing all the way back to the dark ages.

Gamma ray bursts

In an attempt to detect the heat from the first stars, astronomers launched a high-altitude balloon-borne experiment in

2006. It was intended to measure the infrared radiation from the first stars, which had been redshifted into radio waves by the expansion of the Universe (see *How Big is the Universe?*); instead, the experimenters found that a mysterious wall of radio noise deafened their detectors.

This cosmic static was six times louder than anything the astronomers were expecting and completely prevented them from observing the heat from the first stars. Cosmologists speculate that this mysterious radiation may be coming from the death throes of the earliest stars. They have good reason to suspect this because when massive stars explode, they become billions of times brighter than normal. So it seems likely that the first glimpse of something from just after the dark ages will not be an ordinary large star, but the brilliant explosion that marked its death.

Indeed, the most distant object currently known is a type of celestial explosion called a 'gamma ray burst' (GRB). These stellar explosions appear to be more energetic than any supernova in the modern (nearby) Universe and add weight to the idea that the earlier generation of stars were more massive than those of today and so experienced more violent deaths. Each gargantuan star is calculated to have died with a titanic outburst that released as much energy as ten trillion Sun-like stars. Much of this energy was packed into a sudden burst of gamma rays that shot off across the Universe. Watching from Earth, we have no idea where the next gamma ray burst will come from. One arrives every day or two, but from a completely unpredictable direction. Not only that, but astronomers have to be really quick to spot them. Having taken billions of years to travel to us, they arrive and pass by in just a few seconds. Highly sophisticated spacecraft are lying in wait; the instant they detect a burst, they can pinpoint the explosion and, within a second, send out signals to guide other orbiting and ground-based telescopes to look at the correct location.

The gamma ray burst GRB090423 is the record holder for distance, having been calculated to be 13.1 billion light years away when it exploded. Its outburst was detected on Earth on 23 April 2009 and allowed astronomers to calculate that it must have blown up just 600 million years after the Big Bang, making it an excellent candidate for being one of the Universe's earliest stars.

Tuning into hydrogen

Another strategy for investigating the dark ages is to look for the signal from the hydrogen gas that existed throughout space at that time. Hydrogen atoms can spontaneously emit radio waves with a wavelength of 21 centimetres. As that signal travels through the expanding Universe it is stretched to about two metres and can be picked up by an ordinary radio receiver. But in order to process the received data, tens of thousands of radios are needed, all working together, and a supercomputer. Astronomers are now building such 'hydrogen telescopes' and expect to use them to map the distribution of the first celestial objects in the surrounding hydrogen gas. A hydrogen atom can only emit radio waves when it has an electron in orbit around its nucleus; since these atomic electrons would have been stripped away by the ionizing radiation from the first luminous objects, the hydrogen signal would have disappeared at the end of the dark ages. A radio-quiet 'bubble' would have formed around each new celestial object, and the hope is that these will appear as dark holes on the hydrogen maps from the new radio telescopes.

The theory goes that at the centre of each hole on the map will be either a megastar or a black hole. Computer models predict that megastars will 'blow bubbles' in subtly different ways from black holes, and so astronomers believe that they will be able to determine the nature of each central object when they analyse their first results from these telescopes, expected some time around 2015.

Astronomers dream that eventually they will be able to actually see and take pictures of these first objects, and to analyse each object's chemical composition and physical conditions. Only this information will tell them definitively what the first celestial objects were, and how the first wave of heavy chemical elements was generated. Then they can move on to investigating how these individual objects gathered together to become the small early galaxies seen in the Hubble Deep and Ultra Deep Fields. Our current understanding of galaxy formation is still at an early stage: difficulties encountered indicate that either there must be 'dark matter' that we cannot see in the galaxies or our understanding of gravity is wrong.

What is dark matter?

The debate about what holds the Universe together

No one has detected a single particle of dark matter, yet it has become a crucial component of modern astronomical theories. Quite literally, without dark matter, much of cosmology simply falls apart.

The conviction that there must be an extra component of unseen matter in the Universe was founded in the 1930s, when Swiss astronomer Fritz Zwicky was studying the motion of galaxies around one another in the Coma cluster, 320 million light years away. He found that the galaxies were moving so fast that they should burst from the cluster: gravity was simply not strong enough to hold them in. Yet there were many such clusters throughout the Universe and none looked as if they were flying apart, so something was clearly keeping the galaxies together. Zwicky reasoned that there had to be more matter hidden somewhere in the galaxies, providing the extra gravity. He thought that this 'missing mass' must be hiding in titanic clouds of cold hydrogen and helium that had yet to produce any stars. But, try as he might, he could not detect these clouds.

Rotating galaxies

Skip forward 40 years and technology had became sophisticated enough for astronomers to measure not just the speed
at which a galaxy was moving through space, but also how
fast it was spinning. They could even split the galactic rotation into sections and calculate how the orbital speed of
stars varied with distance from the galaxy's centre. The
astronomers were expecting to see stars on the edges of the
galaxy orbiting more slowly than stars near the centre. This
is the pattern displayed by the Solar System's planets, as
discovered by Johannes Kepler (see *Why Do the Planets Stay
in Orbit?*). It is a direct consequence of the fact that gravity
decreases with distance. But it turned out not to be the
pattern followed in most galaxies; instead, no matter how
far stars are located from the galactic centre, they all orbit
with the same speed. When this was discovered in the 1970s,
it presented astronomers with a problem, similar to that of
Zwicky's galaxy clusters, because the outer stars were
moving too fast for the galaxy's gravity to hold on to them.
Since nowhere do we see galaxies in the process of sponta-
neous disintegration, the only solution seemed to be that
there must be more matter concealed somewhere within or
around the galaxies.

For the stars to all move at the same speed, the amount of
this matter would have to increase with distance in a way

that would compensate for the expected drop in the gravitational force. This would require a sphere of matter, called the halo, to surround each galaxy; but while circumstantial evidence for this extra matter was too great to be ignored, the searches for it were finding nothing close to the proportions needed. This discrepancy threatened to plunge cosmology into crisis, until particle physics offered a way out.

Enter dark matter

In the effort to understand and unify the forces of nature, theoretical physicists were contemplating the need for new particles to carry energy. Theoretical predictions of particles had worked well for them in the past; for example, both antiparticles and neutrinos had been predicted to exist before they were observed. Antimatter emerged from calculations made by Paul Dirac in 1928 and was observed four years later (see *How Did the Universe Form?*). Neutrinos were 'invented' by particle physicist Wolfgang Pauli in 1930 to account for missing energy in certain reactions involving the weak nuclear force. He called his invention a 'desperate remedy', but explained that it was necessary either to make up a new particle or to accept that energy could vanish from the Universe. It took 26 years for the experimenters to build an experiment that detected a neutrino.

In light of these successful predictions, particle physicists began talking about whole new swathes of particles in their attempt to unify the fundamental forces. These particles were so unlike normal matter that if they did exist, they would generate gravity but otherwise interact with normal matter hardly at all. Astronomers latched on to this idea immediately, realizing that these theoretical particles might be just what they needed to provide their 'missing mass'. Indeed, when they made their calculations they found that the new matter could outweigh normal matter by anything between ten to a hundred times, making them perfect for providing the extra pull of gravity to hold each galaxy together.

The difficulty was going to be detecting these particles; they were predicted to interact so weakly with normal particles, that there could be giant clouds of them surrounding each galaxy that were completely invisible. Astronomers coined the term 'dark matter' to describe this material and gradually fed it into every cosmological problem that needed more mass. At the same time, physicists were trying to deduce the exact nature of this dark matter. They have offered a multitude of possibilities, and currently it is thought that there may be a pantheon of dark matter particles just as there is a zoo of normal matter particles.

Candidates for dark matter

An early candidate was called the axion, put forward in 1977 to modify the behaviour of the strong force and allow matter to be produced at a slightly higher rate than antimatter (see *How Did the Universe Form?*). Researchers named the particle after a brand of detergent because they wanted to use it to 'clean up' the matter-antimatter problem. In 2005, experimenters thought that they had detected the axion but subsequent investigations disproved this conclusion; the search continues.

In the 1970s, the concept of supersymmetry was propounded by particle physicists. The known fundamental particles can be divided into two categories: fermions and bosons. Fermions are usually associated with matter, because they cannot occupy the same physical space as one another and this resistance leads to matter as we know it. Quarks and electrons are fermions, as are neutrinos. Quarks combine to make the nuclear particles of atoms, which the electrons orbit around. Bosons, on the other hand, are associated with forces, or energy. They can crowd as closely to each other as they like, even occupying the same physical space. Examples include photons, which carry electromagnetic energy, W and Z particles, which carry the weak nuclear force, and the renowned Higgs boson, thought to be responsible for endowing particles with mass. Supersymmetry suggests that every particle has a 'superpartner', thus doubling the expected number of particles in nature. Each fermion has an associated boson, and each boson has an associated fermion. The candidates for dark matter are the bosons' superpartners, which are fermions and so form matter, with mass; collectively these superpartner fermions are called 'neutralinos'.

The Large Hadron Collider in Switzerland may be generating large quantities of neutralinos as it smashes particle beams together, but its instruments cannot detect them. This is because neutralinos are so weakly interacting that they will pass through the walls of the cathedral-sized detectors like ghosts, leaving no trace. This does not mean that they cannot be inferred, however. The more energy that is spirited away by the neutralinos, the bigger will be the deficit when physicists add up the energy of all the particles detected and then compare it to the energy that was pumped into the collision. Any difference between the two could mean that supersymmetric particles are being created but then stealing away from the experiment.

If any energy deficit is seen, it is likely to be a long task to identify which of the many possible species of neutralino is involved. Astronomers cannot help because their computer models deal with clouds of dark matter rather than individual particles. The smallest cloud they can usually deal with is more massive than the Sun, and it is impossible to extract individual particle properties from such large-scale simulations. The only way to know what a neutralino is for sure is to do the almost impossible and capture one.

Capturing dark matter

Although dark matter is expected to interact with normal matter only very weakly, it is supposed to exist in such vast quantities that this would counterbalance its poor interaction capability. At any time there is likely to be so much dark matter passing through a detector that catching a particle is at least feasible. Currently, around a half dozen dark matter experiments are running around the world. Each of them is designed to have the necessary sensitivity to discover a dark matter particle, *if* dark matter exists. Some aim to detect the heat generated when an individual dark matter particle lodges inside the instrument, others look for the dim flash of light created when a dark matter particle ricochets off an atom inside the detector. If any one of the dark matter detectors succeeds, physicists will be able to calculate the properties of the detected particle but, with an expected maximum rate of just a few detections per year, it will take a long time to build up firm results and conclude whether there is more than one type of dark matter.

If neutralinos *are* detected, this does not automatically mean that they will be the dark matter sought by astronomers; they may not exist in adequate numbers or be sufficiently stable to exist for long enough to provide the gravity needed. When it comes to checking the abundance and stability of dark matter, astronomers will need to join the action. Just because supersymmetric particles interact only very reluctantly with normal matter, does not mean that they do not interact with one another. According to models of dark matter, a pair of identical particles will annihilate one another when they collide, releasing a pair of gamma rays. Astronomers expect that such annihilation signals could be coming from the centre of our Galaxy, where the dark matter is expected to be dense, or from neighbouring dwarf galaxies. They are already searching for these signals with satellites.

Once astronomers have determined the relative proportions of axions, neutralinos and any other types of dark matter, they will have to take neutrinos (see *How Did the Universe Form?*) into account because recent experiments have shown that these weakly interacting particles, once thought to be massless, in fact carry a small mass. These would therefore contribute their share of the gravitational pull wherever they are found.

Hidden assumptions

Many astronomers are coming to rely on the existence of dark matter, but others are growing sceptical. This is because dark matter has an Achilles heel in the form of a hidden assumption: its existence relies on Newton's law of gravity being accurate and applicable to gravitational fields no matter how weak they might be. Scientists already know that Newton's law fails in strong gravitational fields, where Einstein's General Theory of Relativity needs to be applied instead. It is possible that the same is true at the other end of the scale, where gravity is extremely weak.

In 1981, as others were beginning to embrace dark matter, Israeli physicist Mordehai Milgrom proposed a change to Newton's law of gravity to explain rotating galaxies without the need for new particles. Remembering that gravity produces acceleration in objects (see *Was Einstein Right?*), Milgrom's suggestion was that Newtonian gravity changes below a certain acceleration value, so that instead of the force dropping as an inverse square law (double the distance, quarter the acceleration), it begins to drop less sharply, as a simple inverse law (double the distance, halve the acceleration). The critical acceleration at which this happens is minuscule – no more than that produced by the gravitational field of a single sheet of paper – but it has a dramatic effect at the outer edge of galaxies.

Milgrom showed that by making weak gravitational fields pull a little harder than expected, theoretical models could successfully show the stars moving with uniform speed all the way out to the extremities of a galaxy. In the vast majority of cases, this produced better agreement with observations than the dark matter models. He called his idea 'Modified Newtonian Dynamics' (MOND) and, although it currently remains a minority view, if the dark matter detection experiments fail to produce any results, and neutralinos are not produced at the Large Hadron Collider, then perhaps more and more astronomers will contemplate what once seemed impossible: that Newton's theory of gravity needs an update.

The drawback that some see with MOND is that it has no theoretical underpinning. This makes many astronomers reluctant to take it seriously, but there are historical precedents for this kind of 'discovery first, understanding later'. Kepler's laws of planetary motion had no theoretical underpinning when he first proposed them early in the 17th century. He simply scrutinised the data and found an equation that reproduced them. Only in 1687 did Newton's Theory of Universal Gravitation provide an understanding of why Kepler's laws worked. Throughout the history of astronomy, laws have been deduced from observations before theories could explain them.

Yet MOND is not flawless; the hypothesis struggles to reproduce the motion of clusters of galaxies, requiring more matter than can be seen to make things work. So are we back to needing exotic dark matter? Possibly not, as astronomers think that there may be a substantial amount of normal matter hiding in galaxy clusters, in the form of warm gas. If they can detect this, using telescopes sensitive to the ultraviolet light it is believed to emit, it would solve MOND's difficulties with galaxy clusters.

The battle continues

In their quest to prove the existence of dark matter, astronomers have recently been using a technique called 'weak gravitational lensing'. This idea comes from general relativity's concept of curved space (see *Was Einstein Right?*). They find a nearby galaxy cluster that would create a large curvature in the fabric of space, and then look through the cluster at the light from more distant objects. As this light passes through the cluster, it would be deflected by the curvature and this would distort the appearance of those distant objects, rather like the celestial equivalent of a hall of mirrors. By charting these distortions, astronomers hope to map out the curvature of space around the galaxy cluster. If there is more curvature than can be explained by the distribution of the galaxies, then one solution is to invoke dark matter, and use the curvature to construct a map of the dark material. Another explanation, however, is to say that the extra curvature shows where the MOND correction is at work, producing more curving than expected.

A particular battleground for the competing theories is the Bullet Cluster, where two galaxy clusters are colliding.

Observations show that the gas between the galaxies has been stripped away, and yet weak lensing indicates that a strong curvature still persists around the galaxies. This can be interpreted as either the gigantic halos of dark matter needed to hold both clusters together, or as regions where MOND needs to be applied. Other galaxy cluster collisions show similar behaviour, and likewise do not help astronomers decide between the dark matter and MOND theories. But very recently something has been found that may break the stalemate, an object that perplexes everyone: a most peculiar galaxy cluster collision known as Abell 520.

In this particular cosmic smash-up, weak lensing results imply that somehow the dark matter has been separated from the galaxies. This is completely contrary to what dark matter theory says should happen, and the only way to explain the dark matter behaving like this is to propose that it is interacting with itself, producing a force that only dark matter feels. Such a hypothesis would solve a problem with the Bullet Cluster collision, as well: computer models cannot substantiate the speed at which the clusters are crashing together, but a previously unanticipated dark matter force could provide the extra kick needed.

At present, the picture is more confusing than it has ever been. MOND does not work in all situations: it needs about twice the amount of matter that we see in the Universe; and dark matter needs a new, exclusive force to make its figures tally. So the bottom line is that no one knows what dark matter is, or whether it even exists. It may be an illusion brought about by our misunderstanding of gravity. In the next few years, the welter of new experiments – the Large Hadron Collider, the dark matter detection devices, and the satellites poised to receive dark matter annihilation signals – promise new information that will surely reveal dark matter to us or disprove it once and for all.

What is dark energy?
The most mysterious substance in the universe

Recent observations have led to the idea that the Universe is filled with a mysterious new substance, a substance so overwhelming that it accounts for three quarters of all the mass and energy found in space. Astronomers have called it 'dark energy' for want of any idea about what it really is. There is no natural explanation for it in any current theory in physics.

'A long time ago, in a galaxy far, far away . . .'. This may not be the most original way to open a chapter (any self-respecting *Star Wars* fan will be sure to recognize it), but for astronomers it has a special resonance. A long time ago, in a galaxy far, far away, a star became a supernova. The light from this explosion travelled for billions of years through space before a tiny fraction of it fell into telescopes on Earth during the late 1990s, and changed the way we think of the Universe. At first, the observation just seemed a little odd because the supernova was somewhat dimmer than astronomers expected. They paid it little attention, reasoning that perhaps something was wrong with their measurements. But as more very distant supernovae were observed, so the discrepancy repeated itself, and what started as a scientific curio turned into one of the greatest challenges to our modern understanding of the Universe.

The supernovae that revealed dark energy were being studied by two independent teams of astronomers who were using them to measure the expansion rate of the early Universe. They then planned to compare those rates with the well-measured expansion rate for today's Universe, and calculate by how much this expansion had slowed down. All astronomers believed that the expansion rate was greater in the past, because the gravity of the celestial objects would be slowing it down. But, when the teams completed their calculations, they had a great surprise: the rate of expansion in the past was *lower* than it is today. Far from slowing down, the Universe's expansion is accelerating. As the researchers investigated further, their observations led to the conclusion that some form of 'antigravity force' had switched on throughout space, and was now driving the celestial objects apart. The prestigious academic journal *Science* named the discovery its 'Breakthrough of the Year' in 1998. Once astronomers became accustomed to the new idea, they welcomed this accelerating Universe because it offered a potential solution to an accumulating body of nonsensical observations.

Troubling observations

The 1990s were dichotomous years for cosmology. On the one hand, the theory of the Big Bang was no longer seriously doubted, yet, on the other hand, puzzling observations were beginning to strike at the very fundamentals of our understanding of the cosmos. There was the age crisis (see *How Old is the Universe?*), in that some celestial objects had been measured to be older than the Universe itself – clearly an impossible situation. Then there were the inventories of matter: studies of the microwave background radiation were telling astronomers that the distribution of matter and energy, just 380,000 years after the Big Bang, was a finely balanced mix that produced no overall curvature of space-time, a situation known as a 'flat Universe' (see *How Did the Universe Form?*).

The mass density for a flat Universe corresponds to an average of just 10^{-26} kilograms per cubic metre; although this might sound minuscule, astronomers consistently failed to reach

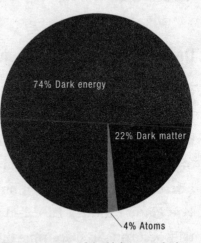

74% Dark energy

22% Dark matter

4% Atoms

The components of the Universe: calculations suggest that almost three quarters of the Universe is dark energy, most of the rest is dark matter and just a small percentage is matter as we know it

this figure when they added up everything they could see in the Universe. Detailed analysis of the microwave background readings, this time taken from spacecraft, confirmed the mass deficiency. The abundance of mass was just four percent of that needed for a flat Universe. Even if astronomers included all of the hypothetical dark matter proposed to hold galaxy clusters together and to make galaxies rotate correctly (see *What is Dark Matter?*), they still only reached 26 percent of the required density.

The discovery of the accelerating expansion not only solved the age crisis, by making the Universe appear younger than it actually is, it also gave cosmologists a way to 'flatten' the Universe.

Dark energy equals missing mass

One of the earliest conclusions that Einstein drew from his Special Theory of Relativity was that mass and energy are interchangeable. He investigated, mathematically, the effects of placing energy into space and found that it causes a curvature of the space-time continuum, just as mass does (see *Was Einstein Right?*). He called the amount of energy naturally occurring in space-time the 'cosmological constant'. It exists rather like the energy in a glass of water at room temperature: this energy may not be immediately obvious, but it must be removed before the water can freeze.

Working before Edwin Hubble's discovery of the expanding Universe, Einstein calculated the amount of energy – the value of his cosmological constant – necessary in space-time to resist all the gravity generated by the various celestial objects and prevent the Universe from collapsing. When Einstein learned of Hubble's discovery of expanding space, he considered that his cosmological constant was superfluous and famously called it his biggest blunder. The realization of the accelerating Universe, however, has made astronomers think there may be unseen energy in space-time after all.

If the 'vacuum' of space contains enough energy, it will overcome the force of gravity between the celestial objects and drive the Universe to expand at an ever-accelerating rate. Physicists tend to refer to this as 'vacuum energy', whereas astronomers have taken to calling it dark energy, to underline its mysterious nature. Whilst it may sound similar to dark matter, and indeed a few researchers are trying to find links between the two, most believe they are nothing to do with each other because they work on different scales. Dark *matter* was introduced to solve the movement of galaxies; dark *energy* was invoked to solve the accelerated expansion of the whole Universe.

To account for the observed acceleration, the density of dark energy needs to be only very low, which is probably why we do not see its effects at small cosmological scales, such as within the Solar System or even our whole Galaxy, the Milky Way. Only when a vast swathe of space lies between a distant object and us does the cumulative effect of dark energy become evident. When the amount of it is totalled up across the entire Universe, it becomes over-whelming, accounting for around three-quarters of the mass and energy in the cosmos, and rendering space almost perfectly flat.

Einstein's theory incorporating a cosmological constant as dark energy does not say where this energy comes from; it just computes its effects once it is there. In the years since the discovery of the accelerating Universe, all attempts to explain a positive cosmological constant have run into severe problems. To start with, it is difficult to conjure a cosmological constant from present physics, some of which has even been designed specifically to remove it. While quantum theory actually predicts a colossal cosmological constant: 10^{120} times (one followed by 120 zeros) larger than that inferred by the acceleration of the Universe, supersym-metry theory (see *What is Dark Matter?*) was formulated in order to get this down to zero. With its proposed raft of mirror-image particles it succeeds in cancelling out the cosmological constant altogether. But supersymmetry was developed before the existence of dark energy was proposed; if the dark energy is Einstein's cosmological constant then we will have to either scrap supersymmetry or revise it in a way that no one can currently conceive. Faced with such a choice, scientists have been searching for other dark energy solutions. Their next proposal was 'quintessence', different from the cosmological constant because it is not a vacuum energy but a new long-range force.

A quintessential solution

Physicists have spent a lot of time trying to understand the disparate forces of nature, and have settled on the belief that just four govern behaviour in the Universe: gravity, electromagnetism, the strong nuclear force and the weak nuclear force (see *Was Einstein Right?*). Occasionally it has been suggested a fifth force of nature must also be present, but no conclusive evidence has been found. Now, however, the accelerating expansion gives them ample justification in saying that some new force could be driving the cosmos. Named in honour of the classical fifth element thought by the Greeks to compose the celestial objects, the quintessence force has one property that the cosmological constant does not: variability. Whereas the cosmological constant is the same everywhere, quintessence can in principle vary with time, and from place to place, making it far more versatile. A number of different versions of quintessence theory have been proposed, each depending upon how fast the force varies with time. One version, known as 'tracker quintessence', closely follows the density of matter and energy in the Universe to produce a gradual acceleration with time. The most extreme version is termed 'phantom energy', which builds up inexorably so that the expansion moves faster and faster until eventually the Universe rips itself to pieces (see *What Will Be the Fate of the Universe?*).

Being a force, quintessence is expected to be generated by the celestial objects themselves, just as electromagnetism or indeed gravity is generated. This produces an overall force field that accelerates the Universe on its largest scales. On smaller scales, it will also generate a force between individual celestial objects and hence move them around, just as gravity does. In other words, if quintessence exists, we should see motions that cannot be explained by gravity.

Some anomalous motions have been explained by invoking dark matter or 'modified Newtonian dynamics' (see *What is Dark Matter?*), and a small number of researchers persist in attempts to see whether these motions are what might be expected from quintessence, thus explaining both dark matter effects and dark energy effects as different manifestations of the quintessence force. Most, however, believe that the two phenomena are entirely unrelated, but finding a form of quintessence that can accelerate the Universe while leaving individual celestial objects untouched is proving tricky. So some physicists are thinking that, instead of adding a new force of nature, the solution might be to modify an old one. Perhaps there are unexpected properties of gravity that appear over gargantuan distances, which Einstein's general relativity did not predict.

Decaying gravity

It turns out that modifying gravity is just as difficult as making quintessence work. Modifications to give a large-scale acceleration usually introduce changes on the smaller scale too, which would affect the movements of the planets by noticeable amounts. With the best of our technical expertise, we do not see such effects; planetary movements are accurately explained by Newton's gravity or Einstein's general relativity. However, an American theoretician, Gia Dvali, has developed an imaginative modified theory of gravity called DGP, in which the particles that carry gravity, the hypothetical gravitons, have a small mass and so impart an underlying shape to space-time.

To produce the observed acceleration of the Universe, DGP gravity allows the gravitons to decay with a half-life of 15 billion years (this means that in 15 billion years their number diminishes to half). As the gravitons steadily disappear – ending up in parallel universes (see *Are There Alternative Universes?*) – so the strength of gravity between objects decreases and the Universe's expansion speeds up. This behaviour would alter the Moon's orbit by about a millimetre a year, and the detection of such an amount is just within the capability of current lunar laser-ranging experiments. If they show nothing on this scale then DGP gravity cannot be right.

Walls and voids

Perhaps the most outrageous and yet paradoxically the most conservative explanation for dark energy is to overturn an assumption so ingrained in cosmology that most people have forgotten it is an assumption. It is the 'cosmological principle', which essentially states that viewed on a sufficiently large scale there are no preferred directions or preferred places in the Universe. In astronomical parlance, it is said that the Universe is homogeneous and isotropic, that is possessed of the same composition and properties, and uniform, in all directions. Alexander Friedman introduced the cosmological principle in 1923 to make it possible to solve the equations of general relativity. Using this assumption meant that Friedman could think of matter as a uniform fluid filling space, and as such it allowed the early cosmologists to investigate relativity and to predict the Big Bang. Since that time, cosmologists have clung to the idea, despite finding ever-larger density variations across the Universe.

The awareness of larger structures in the Universe began around 1937 when Clyde Tombaugh, who found Pluto, studied 30,000 galaxies and discovered that many were found in clusters. He also saw giant voids that appeared to be empty of galaxies. Taken together, the clusters and the voids provide density variations that are not compatible with the cosmological principle, unless one looks at larger scales so that these density variations can be averaged out. The further astronomers looked, however, the more structure they seemed to find, and we now know that clusters of galaxies are gathered together to form superclusters that spread across space for hundreds of millions of light years.

The first example of such a gigantic structure was seen in 1989. The so-called 'Great Wall' of galaxies is some 500 million light years long and 200 million light years wide, yet a mere 15 million light years in depth. This spurred a tremendous period of surveying in the 1990s, when astronomers used ever more powerful instruments to simultaneously record the light from dozens and even hundreds of galaxies. They made hundreds of thousands of observations, enough to give a reliable picture of how galaxies are distributed across the Universe, and found that only on scales greater than 300 million light years could the Universe be said to obey the cosmological principle. And newer discoveries cast doubt even on this. In 2003, another great wall of galaxies was found, a much greater wall, in fact, because it stretches over 1.37 billion light years in space.

Even more recently, there was the hint of a gigantic void in space, seen in microwave background data as a cold region about a billion light years in diameter. Radio telescope observations subsequently confirmed the dearth of galaxies in this vast tract of space, some 40 times larger than any previously known void.

There is currently no explanation for these observations. If we abandon the cosmological principle entirely and admit that the Universe cannot be thought of as more-or-less the same everywhere, then some effects suggested by general relativity that are usually considered negligible would become increasingly important. At the very least, we should expect the speed of the expansion of galaxies to be different in different places. It is thought that we may be located in a low-density region of the Universe; if so, then the Universe might indeed appear to be accelerating because our local bubble would be expanding slightly faster than the surrounding regions.

The trouble is that if the assumption of uniformity is abandoned and we accept that the distribution of matter is more complex than we thought, the mathematics of general relativity becomes impossible to solve. So how do we make progress? The only way to choose between the four options: cosmological constant, quintessence, modified gravity or complexity, is to conduct bigger and better surveys of galaxies.

Eyes on the sky

Surveys are not usually thought of as being the most exciting science that can be performed. Mostly, they are necessary preparation in the quest to find interesting objects for more detailed study. But in the search for dark energy, surveys are essential because the galaxies are like leaves afloat on the cosmic ocean. They are moved around both by gravitational forces and by the expansion of the Universe. The further we look into space, the more the expansion overwhelms the individual movements – and it is this movement of the cosmic 'ocean' we are interested in seeing.

Systematic observations of the furthest reaches of space will allow astronomers to measure the acceleration of the expansion at different times in the history of the Universe. Comparing these accelerations will tell them whether the acceleration has speeded up, or remained constant, or was maybe suddenly turned on at some critical moment. This will hopefully reveal the nature of the Universe's mysterious dominant component that today we call dark energy.

Are we made from stardust?
The mystery of how life emerged

The chances are that you are wearing a gold ring or some other adornment fashioned from a precious metal. Take a good look at it; the atoms in that piece of jewellery are older than the entire Earth. Then consider the iron in your blood, the calcium in your bones, the oxygen that you breathe – all those atoms are older than our home planet and all were forged inside a massive star. Understanding how this star-dust has been processed into living organisms is one of the thorniest questions in science.

Trapped within the meteorites that fall to Earth is a wealth of clues to our origins as stardust. Planetary scientists have even found miniscule samples of stardust itself, which they call 'pre-solar grains'. The tell-tale sign may be a silicon carbide crystal about a millionth of a metre across, or a nano-diamond containing just a hundred or so atoms. When meteorite material is analysed, stardust stands out because of its unique mix of isotopes (elements of a particular mass) – the tiny crystals have remained intact since they were formed, either in an expanding cloud of supernova debris or in a red giant star's tenuous atmosphere. The pre-solar grains grow in these regions because of the relatively slow flow of warm gas there, rather like soot forming in chimneys. The constituents of the grains allow scientists to work out what processes take place inside stars to produce the chemical elements of the world today.

All the elements we find on Earth, with the possible exception of hydrogen, were created by nucleosynthesis (nuclear fusion reactions) inside the cores of stars (see *What Are Stars Made From?*). Somehow, at some time in the Earth's history, the chemicals of stardust have grouped themselves together in such a way that living systems have evolved. This process could be said to be the ultimate mystery: how life formed.

The CHNOPS recipe

Each step, from inorganic chemicals born in stars to living cells, involves crossing boundaries between the three traditional sciences of physics, chemistry and biology. Across each boundary, matter begins to behave differently and manifest unanticipated properties. For example, put enough subatomic particles together and they organize themselves into atoms based on the laws of physics. As soon as these atoms start to interact with one another, physics hands over to chemistry because the panoply of chemical interactions is difficult, if not impossible, to predict from the laws of physics. When a sufficiently complicated network of chemicals comes together, life spontaneously emerges and chemistry can no longer predict the rich variety of behaviours.

To tackle the problem of how life began on Earth, we need to go back to the final stages of the planet's formation, when it was pummelled with asteroids and comets (see *How Did the Earth Form?*). This 'late bombardment' began 4.6 billion years ago and lasted approximately 700 million years. It brought water and other volatile materials to the planets. (In this context, volatile describes chemicals that vaporize at relatively low temperatures, such as water, carbon dioxide, methane and ammonia.) All of the elements contained in the molecules of these volatile substances are from the so-called 'CHNOPS' range of elements, upon which life on Earth is based: carbon, hydrogen, nitrogen, oxygen, phosphorus and sulphur. Of these, oxygen happens to be the most abundant element on Earth, making up nearly half the mass of our planet, with most of it bound into rocks rather than found in the atmosphere. Similarly phosphorus and sulphur are found in the rocks from which Earth is made. In other words, the late bombardment brought many of the vital ingredients for the subsequent development of life on Earth.

Darwin's pond

In 1871, Charles Darwin wrote a letter in which he described life's origin as taking place in a 'warm little pond, with all sorts of ammonia and phosphoric salts, lights, heat, electricity, etc. present, so that a protein compound was chemically formed ready to undergo still more complex changes'. Our current knowledge, however, suggests that this gentle scenario may be a long way from the truth.

The late bombardment created hellish conditions on the Earth. It melted the crust of the planet, threw molten rock into the atmosphere, and evaporated the fledgling oceans. Nothing could seem more opposed to the development of fragile biological molecules, yet intriguingly the evidence suggests that life began soon after the late bombardment began to tail off. There was no sudden end to the bombardment, but a gradual reduction of collisions over time, and by 3.9 billion years ago the impacts had dwindled so much that the late bombardment was effectively over. Scientists find the first evidence for life in rocks dating from just 100 million years after the bombardment stopped. The indication is an enrichment of the lightest stable carbon isotope, carbon-12, which life uses in preference to its heavier cousins because the lighter variety can pass more easily through cell membranes. Wherever living creatures have died, they tend to leave behind a cache of carbon-12.

So, finding the isotope in those ancient rocks has been taken as a sign that microbes were present on Earth relatively soon after the fury of the late bombardment had ceased.

The first fossils are found in Earth rocks dating back 3.5 billion years ago, in western Australia; these are micro-fossils, which are the preserved remains of ancient bacteria, the pre-historic equivalent of pond scum. It may not be a particularly satisfying thought that those were the first life forms on Earth but it seems to be the hand nature has dealt us. The 'cyanobacteria', as they are called, are found as stro-matolites, pillow-shaped communities of bacteria that grow rather like a coral reef. They are formed in shallow water as the cyanobacteria trap nutrient-rich sediments and become cemented into a colony.

So, whatever triggered life on Earth undeniably happened relatively shortly after our planet's formation and the late bombardment subsided. One other fact seems certain. Today, the only way in which life is created is through biological reproduction; it is not spontaneously forming around us, which strongly suggests that the conditions life first formed under must have been utterly different from those that help it thrive today. Darwin recognized this as a problem for his 'warm little pond' hypothesis, which blindly assumed that Earth's environment had always been largely the same throughout its existence. He suggested that the chemical steps towards life are still being taken in the ponds of today, but that anything produced would be instantly palatable and so eaten before it could develop any further. Modern molecular analysis proves that this is not the case, but it does nevertheless suggest other possible routes to life.

Life in a bottle

In the 1950s, chemists Harold Urey and Stanley Miller conducted an experiment that tried to simulate the early Earth. They based their work on the hypothesis of Russian biochemist Alexander Ivanovich Oparin, who in the 1920s, had been the first to suggest that that there is no magical difference between living and non-living matter, that the characteristics of life simply emerge from a sufficiently complex arrangement of matter. He also suggested, spurred on by the discovery of methane in the atmosphere of Jupiter, that methane and its volatile cousins, water and ammonia, were the chemical ingredients from which life formed.

Urey and Miller set out to test Oparin's ideas. They filled a flask with the chemicals they believed existed in the early Earth's atmosphere – chiefly methane and ammonia – and then applied electrical sparks to simulate lightning. As the electricity caused the gases to react together, longer molecules were formed, which dropped into a small pool of water at the bottom of the flask and formed a tarry substance. Upon analysis, this thick gunk was found to contain amino acids, which are the building blocks of proteins. It seemed, miraculously, as if the first step in the process towards life might have been found.

However, as other scientists built on this work, they came to the conclusion that Earth's early atmosphere was more likely to have been composed principally of carbon dioxide. This was bad news because when the Miller-Urey experiment was re-run with a carbon dioxide atmosphere it was nowhere near as successful at producing amino acids. But just when scientists faced a dead end, a new clue landed in their laps – almost literally.

The Murchison meteorite

It was 28 September 1969, late morning in the quiet town of Murchison in Victoria, Australia. A burning fireball split the sky, broke into three and disappeared from view, leaving a smoke cloud hanging. Many fragments of the meteorite were soon found, totalling more than 100 kilograms, and were identified by visiting academics as a rare form of space rock known as a 'carbonaceous chondrite'. Samples were rushed to NASA for analysis, where scientists were amazed to find more than 90 different amino acids within the meteorite material. This clearly indicated that amino acids were assembled in space and brought to Earth during the late bombardment.

However, it seems that the Earth exploited only a small fraction of these amino acids to create proteins – just 20 amino acids go together in different combinations to make up the millions of different proteins used by life on Earth. So the puzzle is how the multitude of 90 or more amino acids led to a functioning life form using just 20.

It is believed that all life today evolved from a single common ancestor, the first organism to form and presumably a very simple living thing. To understand how it came about, a good place to start investigating is amongst the smallest living things on Earth today: microbes. The bacterium E. coli is the 'laboratory standard'; it is rod-shaped and just three millionths of a metre in length. It contains 4377 genes, as compared with the human genome which contains about 40,000. The genes hold the blueprints for the proteins needed to make the life form function. In the case of humans, genes control everything from musculature to eye colour. The common thread between microbes, humans and all other forms of life on Earth, is that all genes are contained in molecules of DNA, deoxyribonucleic acid.

DNA is a long molecule, the backbone of which is a chain of carbon atoms. From this carbon spine hang the genes, each composed of a sequence of chemicals. There are two complementary strands that wrap around each other to form the famous double helix, locking the genes inside. But at specific times the strands can unwrap and make copies of themselves; this ability to replicate lies at the heart of all living things. Yet the replication is not a perfect process; errors creep into the genes when they are being copied and whilst this might seem like a bad thing, it is actually what drives evolution. Although the errors, known as 'mutations', mostly make the proteins behave less successfully, occasionally they improve their function and the life form flourishes, increasing the chances that the favourable mutation will be passed on. By charting mutations and by showing when they diverged from one another, biologists can build up a tree of life that shows how organisms are related. Human beings and other mammals lie near the top of the tree, above simpler animals such as reptiles and insects. Towards the base of the tree are the microbes such as E. coli, and sitting closer to the bottom of the tree than anything else is a group of microbes called the hyperthermophiles.

Black smokers

Hyperthermophiles live on the sea floor in the scalding water surrounding volcanic vents; they are able to withstand temperatures up to 121 degrees Celsius and would actually die if placed in water at less than 90 degrees. They are a subset of a group called the extremophiles, which all live in conditions that would be detrimental to the vast majority of Earthly life. Some extremophiles like acidic conditions or highly salty ones; some like it hot, others like it very cold. They are the ultimate niche organisms. Each variety has evolved a highly unusual metabolism that can extract energy from its chosen extreme environment.

All animal life on Earth requires oxygen, yet various extremophiles can live off hydrogen, iron or many other chemicals that would spell certain death for most living things. Nowhere on Earth contains more readily available chemicals than the volcanic vents on the sea floor, where they are dissolved in the hot water gushing up into the ocean. This makes each volcanic vent a haven for extremophiles. Sea-floor volcanic vents are the underwater equivalent of geysers, such as Old Faithful in Yellowstone National Park. Often known as 'black smokers' because the dissolved minerals condense into black clouds as soon as they hit the frigid water surrounding them, they were first found near the Galapagos Islands in 1977. Most surprising of all, the analysis to place their organisms on the tree of life revealed the hyper-thermophiles to be the most ancient forms of life on the planet.

This could mean that black smokers are the actual sites for the origin of life; they certainly offer some advantages because the hyperthermophiles are the only known colonies on Earth that do not rely on sunlight for energy. If the Sun went out tomorrow, the communities around the black smokers would continue to thrive. Their energy comes from volcanic activity, driven by radioactivity within the Earth, and so they live quite independently of the Sun. This makes them immune to almost anything going on at the surface: ice ages or other climate catastrophes, even the last years of the late bombardment could have taken place without threatening them.

But there is one inconvenient fact that casts doubt on the black smokers as the site of life's origin. The 'tree of life' analysis shows that the hyperthermophiles are indeed ancient but cannot be the common ancestor of all life on Earth. The microbes that evolved into us split away before the hyperthermophiles developed; this is indicated by the fact that hyperthermophiles contain genes that we do not. So, life may have developed elsewhere and some form of it then migrated to the black smokers. We have not yet been able to discover the earliest life forms, either because they are extinct and have left no traces, or because we have yet to look in the right place. This has led some scientists to think that life's original form may be even less complicated than a microbe.

Nanobes

In 1996, geologist Philippa Uwins discovered tiny growths on rock samples retrieved from oil wells. The rocks came from between 3 and 5 kilometres (1.8 to 3.1 miles) below the sea floor, and the growths looked strangely organic. They attracted a dye that binds to DNA, further hinting that they could indeed be living, and researchers began to wonder whether these could be the simplest forms of life on Earth.

They were dubbed 'nanobes' because the smallest ones are just 20 billionths of a metre across. Even the largest are just one-tenth the size of a microbe, and this presents a fundamental problem: there does not appear to be enough space within a nanobe to hold the DNA copying machinery usually found in a microbe. Unless nanobes contain some simpler kind of DNA-copying system, they would appear to lack the necessary mechanism to be alive. If the nanobes *are* confirmed to be living, they hint that life began deep inside the Earth. As the nanobes migrated upwards, perhaps they evolved into the microbes we find around the black smokers.

Since nanobes were discovered, other minuscule 'life forms' have been suggested but the question of whether such incredibly small things can truly be alive is a controversial area of study, with no firm conclusion yet reached. How will scientists know for sure that these tiny objects have taken the leap in complexity from non-living to living matter, crossing the boundary from chemistry to biology? At present, science cannot answer this question because it lacks an incontrovertible definition for life.

What is life?

The best route in trying to define life is to list a number of traits that it must have. For example, we could say that all living things must: 1 – eat or take in some form of energy; 2 – excrete what they do not use; 3 – respond to their environment, usually by moving; 4 – reproduce and pass on traits to their progeny; 5 – be capable of having those traits change between generations. So far so good, but then the mule springs to mind. It is the offspring of a male donkey and a female horse, and in the vast majority of cases it is infertile. By strict application of our rules above, the mule fails points 4 and 5, yet it is undoubtedly alive. Now consider a virus: it needs to invade a living cell in order to hijack the copying mechanism and reproduce itself. So its status as a living organism is debatable.

Coming up with a strictly applicable definition probably requires a new way of looking at life. Think about how to define water: we might say 'a clear, colourless liquid'. Unfortunately this will not do, because so is ammonia, to name just one noxious liquid that you would not want to confuse with water. The only way to pin down water precisely is to describe its chemical composition: H_2O. Before we had knowledge of atoms and chemistry, it was impossible to define water – we just knew it when we tasted it. We are in the same situation with defining life. We know it when we see it, we can take a stab at describing it, but as yet we cannot define it.

A way may present itself through a branch of mathematics called 'information theory'. Founded in 1948 at the dawn of the computer age, information theory seeks to quantify information and find the fundamental limits that govern its storage, processing and communication. Since DNA carries information in the form of genes, perhaps if we regard biology as a form of computation we may be able to define life mathematically. Imagine that a biological system is a computer: it takes information from the genes, rather like a computer reading a programme from a hard drive. It similarly processes the information and creates an output. In the case of a living system, this output is expressed in the form of proteins. So, it is credible that the definition of life lies in the way our cells process the information content of our molecules. Certainly there is no active information processing taking place in a rock.

Work continues to investigate this analogy between living systems and computers, in an attempt to see if it can be transformed into a mathematical definition of life. Simultaneously, in laboratories researchers try to find out whether there are simpler molecules that can replicate themselves like DNA. If so, these could have held the genes of earlier, less complex forms of life. Further clues may come from space probes that are being sent to other planets and to comets to look for amino acids and other building blocks of life. Although we can say with absolute certainty that we are made of stardust, how that stardust then transformed itself into life remains a mystery.

Is there life on Mars?
The chances of finding we have neighbours

If Martian life exists, it is likely to be confined to small niches, protected from the worst of the hostile conditions on the surface of the planet. However, Mars may not have always been so inhospitable. Life could have been widespread in the past.

There are few places on Earth as evocative as Mono Lake, California. The glassy surface stretches off into the distance, and eerie rock formations protrude from the water, like gnarled fingers. It sits high in the hills, about 580 kilometres (360 miles) north of Los Angeles, and has been a remarkable witness to the Earth's geological history. With an estimated age of nearly 800,000 years, the lake has seen an ice age come and go, and has survived the volcanic eruptions that created two islands within it. Walking its banks, you would be forgiven for thinking that you had been transported to another world: Mars perhaps. Certainly that is what the legions of visiting astrobiologists – scientists concerned with finding life on other planets – think as they contemplate this mysterious environment. The landscape

The tufas of Mono Lake, California: similar structures may be found on Mars.

is how planetary scientists imagine Mars to have been in the past, before the last of the water was lost from its surface, forcing any Martian life that existed to struggle for survival.

Attracting the most attention are Mono Lake's eerie fingers of limestone, known as 'tufas', protruding some three metres from the surface of the lake. If astrobiologists could find such structures on Mars, these might reveal whether life ever began on the red planet. The tufas formed underwater when calcium-rich spring water bubbled up through the alkaline lake. The lake water was rich in bicarbonate and this combined with the calcium to form limestone, which aggregated to build the tufas. As the water level in the lake fell, so the tufas broke the surface. Crucially, the tufas contain an abundance of microfossils, entombed as the limestone built up. A tufa landscape on Mars would offer a perfect landing site from which to search for Martian microfossils.

Extreme living

The initial attempt to find life on Mars was made in the 1970s by a pair of landers called Viking 1 and 2, the first manmade objects to touch down on the red planet. They carried out four experiments to look for life by analysing the Martian soil, but only one showed any promising results. When a nutrient broth was dripped onto the soil, carbon dioxide gas was liberated, mimicking microbial metabolism. However, subsequent runs failed to repeat the release. Also, the other experiments failed to detect the organic molecules expected in the Martian soil if microbes were present. So astrobiologists concluded that life was not present in the soil sample but this left them with a question: what was it that released the first puff of carbon dioxide? This was to remain unanswered for several decades.

In 2008, NASA's Phoenix robotic lander detected chemicals called perchlorates in the Martian soil, which could have reacted to give off carbon dioxide. They may also be responsible for breaking up any organic molecules in the 'top soil' of Mars. Whatever the exact explanation, the search for life on Mars is proving far more difficult than just landing robots on the planet and scooping up a handful of dust. What has been revealed without doubt is that the conditions on Mars today are extremely hostile by most Earthly standards. Not only are there highly reactive chemicals in the ground, there is no air to breathe, and the surface is scoured by ultraviolet radiation and high-speed particles thrown out by the Sun. There is an apparent lack of water and the temperature swings from around 17 degrees Celsius during the day to lower than minus 100 degrees at night.

In micro-niches on Mars, however, conditions may be right for organisms that would be considered extremophiles if found on Earth (see *Are We Made From Stardust?*). Failing that, it is possible that life might have existed on Mars in the past when conditions were different.

Follow the water

Planetary scientists are certain that Mars began its existence as a world similar to Earth, with many bodies of standing water. If life began on Mars soon after the cessation of the late bombardment (see *How Did the Earth Form?*), as it is believed to have done on Earth, there could have been plenty of time for it to colonize the planet before the conditions changed.

In the quest for life on Mars, the mantra is to 'follow the water'. This is because life needs a liquid in which to perform its chemical reactions. Water is an excellent option because it is a simple molecule and is abundant across the Solar System. Until astrobiologists are forced to consider more outlandish alternatives, they have decided to look for water first and then, once they find it, to look for signs of life in that location.

NASA's Viking orbiters of the 1970s returned pictures that show geological structures on Mars resembling water-cut features on Earth. Some of these, such as the Nanedi Vallis, are indicative that water has run on Mars as rivers. This meandering channel is 2.5 kilometres (1.6 miles) long and displays terraced walls and oxbow curves. It is clearly a place in which water has run repeatedly, perhaps constantly, over long periods of time. Other features, such as the tear-drop-shaped 'islands' in the mouth of Ares Vallis, show that water has flowed across the surface as floods. The islands appear as if they were fashioned by millions of gallons of water emptying from the valley onto the surrounding plains. This picture was confirmed when NASA landed the Pathfinder rover in the outflow plain near the teardrop islands in 1997. Its images were immediately notable because many of the rocks surveyed were rounded, as they would be by erosion in a watery environment.

There is even the suggestion that Mars once had a large ocean. Firstly, there is circumstantial evidence in the difference between the two hemispheres of Mars. The southern hemisphere is mostly high ground, with jagged craters and chasms, whereas the northern hemisphere is a low-lying basin, and it is much smoother, implying an earlier ocean. Secondly, there are two possible shorelines. Seen from orbit, each stretches for thousands of kilometres around the northern basin, and they have been estimated to be between two and four billion years old.

Yet today the planet seems to be dry. If we are going to target areas to look for life, then we need to understand what happened to the water and, if any of it is still on Mars, where it could be hiding.

Where did the water go?

Mars has been badly affected by its lack of a magnetic field. Unlike the Earth, where harmful solar particles are mostly deflected away, these collide with Mars's atmosphere. This has eroded away the gas by knocking molecules out into space, and as the atmosphere was lost, so the planet's climate declined. The decrease in atmospheric pressure meant that lakes, seas, even the ocean would have evaporated more easily. With no protective magnetic field or ozone layer, the incoming solar radiation would have broken up airborne water molecules into their constituent hydrogen and oxygen atoms. The liberated hydrogen would have been light enough to escape the planet's gravity and disappear into space, whereas the heavier oxygen would have sunk to the surface of the planet and combined with the minerals in the rocks. Any water that remained on the surface is thought to have drained downwards into the rocks and frozen solid.

Early theories assumed that this had been a relatively gradual process, perhaps lasting billions of years. However, by counting the number of craters in different regions of Mars, planetary scientists are uncovering a different picture: a past history punctuated by planet-wide climatic and volcanic catastrophes. Counting craters seems simplistic but it is useful because during volcanic eruptions areas of the planet were resurfaced with lava, erasing the existing craters and providing a blank sheet onto which meteorites fell during the subsequent aeons, creating new craters. By comparing the number of craters in each region of Mars, planetary scientists can estimate the ages of those areas: the surfaces with the fewest craters are deemed to be the youngest.

Instead of undergoing a gradual transition over four billion years from an Earth-like world to the present frigid desert, crater counting suggests that Mars became a desert in less than a billion years. Any widespread life that had developed on the planet would have perished, and the remainder would have been driven into niches, much earlier than previously thought. The planet was then subjected to a violent sequence of volcanic activity, producing a series of giant upheavals. Each one briefly resuscitated the planet, because the sudden outpouring of internal heat would have thawed out the frozen reserves of underground water and driven them upwards to the surface, flooding large areas of the planet. According to analysis, there have been five such upheavals during Mars's history: the earliest took place 3.5 billion years ago; this was followed by another 1.5 billion years ago; then by further events 800 million years ago, 200 million years ago, and 100 million years ago. Each episode may not have lasted for more than a few tens of thousands of years but they have all left ample evidence on the surface of the planet. This is not just seen in the form of lava flows but also in the myriad outflow channels, river-beds and even shorelines. Perhaps during these periods, Martian microbes flourished before again being forced to return to their deeply frozen conditions.

Ice fields

Today most of the remaining water is thought to be just below the surface in the form of icy plains, or deeper in underground lakes. In 2002, NASA's Mars Odyssey orbiter showed that ice was indeed buried, perhaps just a few centimetres under the surface, in the northern hemisphere of

Mars. When Phoenix touched down in the middle of these plains on 25 May 2008, it swiftly verified that there were large deposits of ice just below the surface. In the spirit of the 'follow the water' mantra, this has boosted the possibility of finding ongoing life on Mars.

There has been no similar luck in finding the supposed underground lakes of water. Two radar instruments have been sent to Mars, both designed to send radio waves down to a kilometre or more beneath the surface where they would be expected to bounce back from the supposed boundaries between rock layers and water layers. Thus, they should have created a map of Mars's network of underground lakes. However, neither radar instrument has seen anything resembling a lake. Perhaps the water is deeper than anticipated, or simply not there; either way, this particular result is not so encouraging to the search for life.

It is ironic that as planetary scientists collect more and more data from Mars, they find themselves presented with an increasingly confusing picture as to whether the red planet is habitable or not. They press on in the quest because there is so much at stake. Not only would finding the evidence for past or present life on another planet be a tremendous achievement in itself, it could also fill in some important gaps in our knowledge of life on Earth and specifically how life first formed here.

Primordial evidence

On Earth, the fossil record of life's origin has been lost because of our planet's restless surface. Driven by the decay of radioactive elements in the Earth's interior, our continents float on a semi-molten layer, moving by a few centimetres each year in a process called plate tectonics. If the continents rub past one another they produce earthquakes; if they push against each other, they build mountain ranges. If one plate rides roughshod over another, it forces it down into the interior of the Earth, where the rocks are melted and recycled in the form of lava that bursts through volcanoes at the surface.

This global recycling has destroyed nearly all of the truly ancient rocks on Earth, taking with them any traces of the first life forms. This may not be true for Mars; being a smaller planet, there was not enough radioactive heat to begin a full-scale plate tectonics process. Lacking this constant 'recycling machine', parts of Mars – like the Moon – must be primordial, dating back to the original formation of the planet. A meteorite called ALH 84001, discovered near the Allan Hills in Antarctica in 1984, bolstered this belief. Dating ALH 84001 showed that it was around 4.5 billion years old, placing it at the very origin of the Solar System. Small bubbles of gas were trapped inside the rock and researchers examined their content.

Astonishingly, the gas displayed the same composition as the Viking landers had registered for the Martian atmosphere. The rock appeared to have come from Mars.

The story now envisaged for the long life history of the meteorite is that the rock solidified as part of Mars's original surface but was blasted from its site of formation during an impact in the late bombardment, around 4 billion years ago. But this did not loft it into space; instead it fell back to Mars and remained on the surface for nearly the whole of the subsequent history of the Solar System. Then, 13 million years ago, another impact sent it careering into space, where it crossed paths with Earth some 13,000 years ago. It fell as a meteorite, landing in Antarctica, where it was entombed in a glacier and finally returned to the surface as the glacier struck the Allan Hills. And, if that story is not amazing enough, ALH 84001 may even be showing us the very way life began.

Martian microbes?

In mid-summer 1996, NASA showed the world scanning-electron-microscope images of tube-like structures found in a sample of meteorite ALH 84001. They were clearly distinct from the surrounding rock and looked eerily reminiscent of bacteria fossils found on Earth, and chemical evidence suggested that these tubes might indeed once have been alive. But there was no consensus. These 'fossils' were only 20–100 billionths of a metre in diameter, making them similar in size to the nanobes from Earth that were simultaneously being puzzled over (see *Are We Made From Stardust?*). As a result, the same argument against their once having been alive was presented: namely that they were too small to contain the DNA copying mechanisms considered necessary. Of course, the same counter-argument was proposed as well: that these organisms might carry out their copying in a more primitive way, eventually superseded by the reproduction methods of today's microbes. Other scientists claimed that such structures could be made through simple crystallization processes, without the intervention of life.

Another important development in the debate about life on Mars became widely acknowledged by scientists early in 2009. It concerns methane, a gas readily found in Earth's atmosphere, produced either by volcanic activity or by the metabolism of life forms. For several years planetary scientists had been detecting methane on Mars. It is localized to three regions of the planet, rather than spread thinly throughout the whole atmosphere, and this strongly suggests that it is being produced at those locations, otherwise it would have been distributed around the planet by winds. If it is being produced now, then either there is current volcanic activity inside Mars, or there are colonies of living Martian microbes metabolizing under the surface. Either option is quite mind-blowing. If it is life, then the discovery is obviously momentous. But the volcanic option would be big news, too, because Mars was thought to be geologically dead; the volcanic episodes that characterized its past were considered impossible today.

The excitement is compounded because, during follow-up observations, the methane was seen to have disappeared. It had not been blown around the planet; it had totally disappeared. Something had destroyed it. The ultraviolet light from the Sun could not have broken it down that quickly, so scientists think the cause must be highly reactive chemicals in the soil. Calculations show that the methane was destroyed 600 times faster than scientists were expecting, so to have built up the original quantity in the atmosphere requires a production mechanism that worked 600 times faster than previously assumed. If microbes generated it, there must be 600 times more of them than originally thought. But until we can get to Mars and determine hands-on whether it is life or volcanic activity producing the methane, we will remain uncertain of whether there has been, or still is, life on Mars.

Are there other intelligent beings?

Is anyone out there?

The chances of detecting signs of extraterrestrial beings are small, but there is a deep human curiosity about whether we are alone in the Universe and this drives a continuing search.

Even in April, the temperature in Green Bank, West Virginia can be bitterly cold, especially at four o'clock in the morning. Retired professor Frank Drake knows this only too well, as back in 1960, when he was 29 years old, that was the time he started work at the Green Bank Telescope. He was the first human being to conduct a search for extraterrestrial intelligence. He tuned the receiver of the radio telescope to pick up radio waves given out by hydrogen atoms. Given the importance of water (comprising hydrogen and oxygen) to life on Earth, Drake hoped that this fundamental frequency would be a natural carrier for extraterrestrials to transmit signals. He then turned the telescope to the first target: a sun-like star called Tau Ceti just 12 light years away. In the hope that an extraterrestrial signal would come through loud and clear, he rigged a tape recorder and a speaker in the room. But for Tau Ceti, he recorded nothing, so he turned to Epsilon Eridani, the next star on the list. Within minutes the room filled with powerful radio fuzz; he could hardly believe it was this easy. But then the signal disappeared. After days of searching, the transmission returned but this time it was clearly terrestrial interference. Undeterred, Drake and his colleagues continued the search, and are still searching today.

The great silence

Half a century since the search began, there is still no evidence for other intelligent races in our own Galaxy. Some have dubbed this lack of signals 'the great silence'. Yet this is by no means proof of the non-existence of extra-terrestrials. The Galaxy is so large, the radio spectrum so wide, and the technology we have available so limited (compared to what we can imagine building if we had the money), that we have barely scratched the surface in this search.

The best way to understand the magnitude of the under-taking, and the huge uncertainties involved, is to do what Frank Drake himself did in 1960 and try to estimate the number of extraterrestrial civilizations there should be in the Galaxy. Drake attempted this by writing down a long string of factors which multiplied together would give this number:

1: The average number of stars to form per year in the Galaxy.

2: The fraction of those stars that form planets.

3: The fraction of those planets that could support life.

4: The fraction of life-supporting planets that do indeed form life.

5: The fraction of those living planets that develop intelligent life forms.

6: The fraction of those intelligent life forms that develop technology.

7: The average lifetime of a communicating species; in other words, how long a civilization will use radio technology, leaking signals into space for us to hear.

Dishearteningly, the only factor that is known is the first one. Astronomers have shown that the Galaxy gives birth to about seven new stars per year. They are now working on an estimate of the second term, the fraction of stars that form planets. It has always been assumed by astrophysicists that our Solar System is typical and that most stars should form planets. However, testing this assumption has proved tricky because it is incredibly difficult to see a planet around another star. A planet does not emit light and so it is hidden in the glare of its parent star. Therefore taking an image is the equivalent of trying to discern a pinhead held next to a searchlight. Nevertheless, during the last 15 years astronomers have inferred the existence of more than 400 planets around other stars, terming them 'exoplanets'. They have found them by detecting the way each planet's gravity induces a wobble in its parent star (see *Why Do the Planets Stay in Orbit?*).

This method is not sensitive enough to detect smaller planets; it is only suited to revealing planets larger than Earth. For a more complete picture of the number and variety of planets, astronomers are now using a space telescope called Kepler to monitor 100,000 stars for the drop in light caused when a planet slips across in front of it, a celestial alignment known as a 'transit'. Far above the distorting effects of the Earth's atmosphere, Kepler is accurate enough to detect the slight reduction caused by transiting Earth-sized planets. Astronomers believe that this will give them a census of how many stars form planets and so allow them to provide an answer for the second term in the Drake equation. Kepler's survey will also help them determine the third term - how many of those planets might be suitable for life.

Of the more than 400 planets currently inferred to exist around other stars, only one is thought to be habitable. It is called Gliese 581c, and orbits a dim red dwarf star just 20 light years from Earth. The planet is 1.5 times larger than Earth, holds between five and ten times more mass, and generates a gravitational field twice as strong. The planet is either a large rocky world, dubbed a 'super-Earth', or an oceanic planet with some similarity to Uranus or Neptune in our own Solar System. Uranus and Neptune have similar masses to the upper estimate for Gliese 581c and their ices would melt close to a star, transforming the planet into a world with nothing but ocean on its surface. Regardless of whether it is rocky or oceanic, astronomers believe it is habitable because it lies just within its star's 'habitable zone'.

The habitable zone

This is the region surrounding a star in which a planet could be warm enough (but not too warm) to support liquid water on its surface. The zone's distance from the star is determined by the star's temperature, which is a signpost of how much energy it pumps into space. Obviously, the Earth lies within the Sun's habitable zone, which extends from between 0.95 and 1.5 times the distance of the Earth from the Sun.

Gliese 581c orbits a red dwarf star, which is significantly cooler than the Sun and so creates a much smaller habitable zone lying much closer to the surface of the star. By comparison, the Earth is 14 times further away from the Sun than Gliese 581c is from its star. Consequently this exoplanet completes an orbit so quickly that its year lasts just 13 Earth-days. At this proximity, the gravity of the star is so strong that Gliese 581c is trapped into showing just one face to the star, with the result that one side of Gliese 581c is in permanent daylight and the other in permanent night. Calculations show that its surface temperature would allow liquid water and so it must be thought of as habitable. The same cannot be said for any of the other known exoplanets, either because they lie outside their star's habitable zone or because they are thought to be gas giants with no solid surface.

In our own Solar System, Venus, at 0.75 times the Earth's distance from the Sun, is well outside the inner boundary of the habitable zone, but Mars is right on the outer boundary at 1.5 times Earth's distance. These are not encouraging statistics for the Drake equation because it means that out of our eight planets, only Earth and possibly Mars can be considered habitable. If the same pattern is repeated across the Galaxy, then it means that only between one quarter and one eighth of the planets detected will be habitable. The Kepler mission will help us to refine this figure by being sensitive to planets of many sizes and orbital configurations, and astronomers are particularly on the lookout for Earth-sized worlds within the habitable zone of their star.

Some astronomers have pointed out that by focusing exclusively on the habitable zone we may be blinding ourselves to exotic possibilities of life, ones that do not require Earth-like conditions. Also, there may be unexpected places where the temperature is suitable for liquid water. Jupiter's moon, Europa, for example, is well outside the traditional habitable zone and yet tidal effects from Jupiter's powerful gravitational field supply it with enough energy to maintain a global sub-surface ocean of water. And, according to the astrobiologists, wherever there is water, there may also be life (see *Is There Life on Mars?*). Nevertheless, until we understand more about such exotic locations, most researchers take a conservative approach, preferring to underestimate the number of habitable planets than overestimate them.

The chance that life emerges

The next term in the Drake equation is the proportion of those habitable planets that go on to create life. Astronomers around the world are currently designing missions that could reveal living planets to us, not by the radio signals emitted into space but by the way life forms alter the chemical composition of their planet's atmosphere. The widespread presence of life on Earth puts methane and oxygen into the atmosphere. Without constant replenishment from plant and animal metabolisms, the gases would disappear because they react with each other making water and carbon dioxide. So, a clear signal of a living planet would be the presence of oxygen and methane. Indeed, the recent discovery of localized pockets of methane on Mars has ignited the hope that the planet may harbour some life (see *Is There Life on Mars?*).

Astronomers need equipment capable of collecting enough light from the feebly dim exoplanets to perform a spectral analysis (see *What Are Stars Made From?*) of their atmospheres. It is a difficult undertaking; only four exoplanets have ever been imaged from Earth: three around a star called HR8799 and one around the bright star Fomalhaut. Once the glare of each central star was blocked out, the exoplanets appeared as tiny points of light, and to stand any chance of chemically analysing these worlds a telescope would have to collect their light for weeks or even months. This is impractical when so much other science could be achieved in that time, so astronomers and engineers are planning to build a dedicated space telescope.

Until this plan is put into action and data is collected, scientists find themselves able to do little more than guess where life may be. It is impossible to say how easily life could have formed elsewhere when we are still unclear about the steps taken for life to form on Earth (see *Are We Made From Stardust?*).

Estimates or guesswork?

The values of the remaining terms in Drake's equation are similarly little more than guesswork, because the only example we have of a living planet is the Earth and no scientist would draw a conclusion from a sample of one.

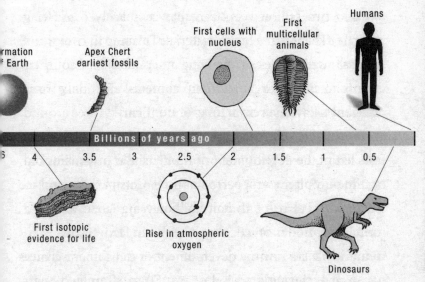

Billions of years ago

4   3.5   3   2.5   2   1.5   1   0.5

rmation
Earth

Apex Chert
earliest fossils

First cells with
nucleus

First
multicellular
animals

Humans

First isotopic
evidence for life

Rise in atmospheric
oxygen

Dinosaurs

Timeline of life on Earth

Nevertheless, it is all we have to go on for the moment. The next two terms are the number of living planets that develop intelligent life, and the number of those that develop technology capable of sending radio messages into space. If Earth's history is typical, those chances might be rather small as there have been many evolutionary hurdles for life to jump before arriving at intelligence.

The fossil record shows that evolution has been a slow but accelerating process on our planet. For more than half the Earth's age, life was confined to the simplest form of cell, known as a 'prokaryote', which lacked a nucleus. Everything that the cell needed to function was simply jumbled together inside the cell membrane. It was only around two billion years ago that cells evolved nuclei in which to keep their genetic material. This leap in complexity was followed by the cells evolving other specialized compartments to better organize their contents and make more efficient use of the resources around them.

Eventually, cells somehow evolved to work together and this led to the development of multicellular organisms but, again, only after a vast period of time. Complex life became successful on Earth only half a billion years ago, in an evolutionary episode called the 'Cambrian Explosion'. This remarkable era saw the development of most major groups of animals during a period of just 70 to 80 million years. The extraordinary rate of evolution may have been stimulated by the build-up of oxygen in the atmosphere, providing more energy to power more complex life forms. The oxygen build-up was actually the worst case of planetary pollution in the history of the Earth. Oxygen is the waste gas given out by photosynthesizing cells, which turn sunlight into their energy for life; it is also the waste product of certain microbial metabolisms. It is a highly reactive gas, and in some circumstances it is toxic because of the aggressive way it attacks biological molecules. As oxygen accumulated in the atmosphere, it triggered a mass extinction of Earth's early microbes, ironically destroyed by their own pollution.

Some ancient microbes survived by finding niches away from the oxygen, for example underground; other species evolved coping strategies. These included binding the oxygen into molecules such as collagen, which then provided structural support so that organisms could grow larger. They also learned how to use the oxygen to generate energy and this proved to be highly successful, leading to all animals breathing oxygen. It is often said that the availability of energy-laden oxygen led to the development of our power-hungry brains. Even so, humans with their intelligent brains did not arrive on the scene until just 200 thousand years ago, and the technological ability to send messages into space not until just 70 years ago. In other words, when taken over the entirety of Earth's lifespan, humans have been around for the merest blink of an eye, and the ability to send radio signals into space for a fraction of that.

It has to be considered that just because a planet can support intelligent life does not mean that those life forms are guaranteed to develop the necessary communication technology. Take, for example, Gliese 581c, the only habitable planet so far known around another star. It is possible that this is a water world, with an all-encompassing ocean and no continents. If so, then it could evolve an intelligent aquatic species but it would seem unlikely that such an underwater civilization would develop electrical technology.

The final term in the Drake equation is the average time for which a communicating civilization exists. It has to be noted that the abilities to transmit signals and to destroy ourselves with atomic weapons arrived at about the same time in human history. This has led some to speculate that intelligent races will not live for long after developing technology, perhaps only a century or two. It does not have to be war that silences us; our technology is undoubtedly contributing to climate change, which could have catastrophic repercussions. Others of a more optimistic outlook believe that we will overcome these problems. If so, then our species and our technology could exist for a very long time. It is estimated that the Earth will remain habitable for some billion more years before the Sun heats up so much that it boils away our oceans (see *What Will Be the Fate of the Universe?*). So potentially, humankind could be transmitting into space for a long time. The 'bottom line' for the Drake equation is that the final term is usually the one that dominates all the rest. The longer we estimate for the lifetime of a technologically advanced civilization, the more of them we should expect to be populating the Galaxy, increasing our chances of eavesdropping on them.

SETI's great dream

Following on from Drake's early attempts, the current era of search for extraterrestrial intelligence (SETI) was provisionally launched in 1971 when NASA commissioned a study into the design for the ultimate SETI telescope. Called

'Project Cyclops', this gargantuan field of interlinked telescopes would have rendered its metaphorical single eye capable of detecting any stray radio communications from planets within 1000 light years of Earth. (This would be the same kind of radio radiation that spills from Earth into space as a result of our radio and television broadcasts and military radar signals.) Traditional radio telescopes are simply incapable of picking up such weak transmissions.

Cyclops was, however, never built and without it SETI scientists have to rely on extraterrestrials targeting Earth with a concentrated radio signal designed to attract attention; this assumption has drawn criticism. Sceptics ask why an advanced alien culture would want to talk to us; it is, they say, the equivalent of our trying to talk to an amoeba. Notwithstanding the scepticism, the widespread interest shown by people in whether we have alien counterparts is underlined by the popularity of the 'SETI@home' programme. Released in 1999, this is computer software that uses home computers' idle time to analyse data from Berkeley's SERENDIP project, which stands for the Search for Extraterrestrial Radio Emissions from Nearby Developed Intelligent Populations. The SERENDIP receiver sits on a radio telescope and collects data from whatever the telescope happens to be observing. The data is then distributed across the Internet to people running the SETI@home software, which analyses it for possible signals and sends the results back to Berkeley. To date some curious signals have been discovered but nothing that has stood up to further scrutiny.

The SETI scientists may find an extraterrestrial signal tomorrow, or next year, or it may be in the next decade or the next century, or never. We currently do not possess the costly technology required to make a definitive search. Given the factors involved, especially the long time it took for intelligent life to develop on the Earth, many astronomers believe that, whilst there may be many planets that develop life in some form, only a very small proportion will go on to evolve intelligent beings with interstellar communication technology. Without any evidence to the contrary, it remains possible that we are the only intelligent species in the Galaxy – or even in the Universe.

Can we travel through time and space?

The possibility of warp drives and time travel

Travelling through the vast distances of interstellar space currently seems impossible, and time travelling is purely fictional. Undreamt of technologies in the future could change all this – but if so, why haven't we met our future selves?

Someone was stealing trashcans. New Yorkers did not see them go, but during the spring and summer of 1950, metal bins were being spirited away from Manhattan's streets at an alarming rate. At the same time, a rash of flying saucer sightings was gripping the country, and a cartoon appeared in *The New Yorker* portraying a fleet of aliens returning to their home planet with their cache of dustbins.

'It's a neat solution to both phenomena,' said physicist Enrico Fermi, when he heard a description of the cartoon one summer lunchtime at Los Alamos National Laboratory, New Mexico, and it led him into a discussion with colleagues about the possibility of faster-than-light interstellar travel.

Einstein's Special Theory of Relativity states that the speed of light is an absolute speed limit throughout the Universe. Nothing can travel through space faster than light and this stands in the way of interstellar spaceflight, because the stars are so far away from one another. The record holder for the fastest manmade object is the spacecraft Helios 2; launched in 1976, it reached speeds of about 250,000 kilometres per hour (155,000 miles per hour) during a series of close fly-bys of the Sun. However, compare this with the speed of light, approximately 1.1 billion kilometres per hour (0.7 billion miles per hour), and it looks like a snail's pace. Travelling at the maximum speed of Helios 2, it would take 18,500 years to reach even our nearest star.

Even if we could travel at the speed of light, there are only eleven stars (excluding the Sun) within a distance of ten light years. It would take 4.3 years travelling at the speed of light to reach the nearest star, a triple star system known as Alpha Centauri, 4.3 light years away. Confined to speeds below that of light, the rest of the Galaxy will forever remain beyond our reach.

Imposing a speed limit

The universal speed limit came about as a consequence of the principle that the speed of light is a constant. By this Einstein meant that regardless of the speed of the measurer, the speed of light always appeared to be the same. This is contrary to what was expected, because in our normal experience velocities 'add' together; two cars travelling at 50 kilometres per hour towards each other pass with a combined, or relative, velocity of 100 kilometres per hour. This is not true for light. If you measure the speed of light the value will always be the same, no matter whether you approach the light beam head-on, from the side, or are running away from it. This had been experimentally proven by the Michelson–Morley experiment in 1887.

American physicists Albert Michelson and Edward Morley originally designed their experiment to provide incontrovertible evidence for the ether. At the time it was widely believed that light needed a medium to travel through, rather like sound travelling through air, and the postulated medium in which the Earth was immersed was the ether. Michelson and Morley reasoned that as Earth travelled around its orbit at 30 kilometres per second (19 miles per second) it should experience an ethereal wind. The wind would sweep across Earth in different directions and at different speeds, depending on the time of year and the direction in which the Earth was moving. To measure the effects of this wind, Michelson and Morley split a beam of light into two using a semi-silvered mirror so that half of the light passed straight through whilst the other half was reflected. They then sent these identical beams down two paths at right angles to one another, and eventually combined the beams on their return journey. Travelling in different directions relative to the Earth's motion, the two light beams should have been affected differently by the etheral wind: one beam would meet it head-on, the other would feel it broadside. However, the light beams were not affected in any discernible way. No matter when the experiment was conducted, the same null result was obtained: it was as if the ether did not exist at all.

Eventually this is what the physics community decided: there is no ether. Light does not need a physical medium, and its perceived speed does not depend on the speed of the observer. Accepting this result, Albert Einstein set about investigating what consequences this would have, and it led him to the Special Theory of Relativity. 'Relativity' here refers to the fact that speeds can only ever be measured relative to something else. There is no universal standard, no absolute framework of space that speeds are measured against; one object must always be compared to another. 'Special' refers to the fact that this is not a result that can be easily applied to all forms of motion; at first, Einstein investigated only non-accelerated motion, in other words the simple case where objects were not changing their speed or direction in any way. He subsequently extended his investigation to accelerated forms of motion in his General Theory of Relativity (see *Was Einstein Right?*).

While working on special relativity, Einstein found that once an object's velocity exceeds ten percent of the speed of light, previously unimaginable effects manifest themselves. As bizarre as it may seem, the object becomes more massive the faster it travels. This increase in mass means an increase in inertia: more energy is required to make the object go faster. So, as the object accelerates further, becoming more massive all the time, greater and greater amounts of energy would be needed; in fact to accelerate an object to the speed of light would require an infinite amount of energy. It follows that it is impossible to attain the speed of light: it is a fundamental speed limit. On the face of it, we seem restricted forever to meander around the confines of the Solar System like ants on Earth's surface, restricted to a minuscule area because we simply do not live long enough to cross the vast distances between the stars. It is a frustrating thought – but there is a get-out clause.

Special relativity tells us that nothing can travel through space at more than the speed of light, but the theory does not prevent space itself from expanding or contracting faster than the speed of light. This is exploited in the theory of inflation (see *How Did the Universe Form?*), which postulates that the early Universe underwent a sudden period of extreme expansion. The inflating Universe flung itself outwards, distributing matter and energy at speeds far faster than light. Even today distant galaxies are moving away from us faster than the speed of light because of the expanding space between us and them. But as they are not actually moving through space at such speeds, Einstein's laws are not broken. This is the weird world of relativity, and astronomers have to get used to it or the observed motions in the Universe make no sense.

Einstein's predictions are now verified on a daily basis. Without relativity we would not have global positioning systems (GPS). Only by taking into account special and general relativity can GPS devices provide accurate locations. In other words, every time you check your position on your mobile or your satnav, you are relying on Einstein's theories. But applications more much exciting than satnavs can be envisaged.

Introducing warp speed and wormholes

Relativity allows us to imagine the basic concept of a Star Trek-like 'warp drive'. Think of it as an engine that can stretch space like a piece of elastic, with your spacecraft being carried along in the stretching. Once you have reached your destination, you move off to one side and let space return to normal behind you. Making this work seemed pure science fiction until theoretical physicist Miguel Alcubierre Moya mathematically modelled a way of forcing space into a configuration that would create a bubble of warped space-time, rather than an elastic strip, which could travel through the Universe. The trick is to arrange an energy field around a spacecraft so that space behind it is forced to expand and space in front of it is forced to contract. Between these two warped regions sits a bubble of flat space-time in which a spacecraft could safely rest and travel at 'warp speed'.

The difficulty with the warp drive model is that to create a contracting region of space-time requires a hypothetical form of matter called exotic matter. Exotic matter, unlike antimatter, has negative mass; this means that it would feel repulsion in a gravitational field. Although there are hints of 'exotic' behaviour in the Universe, for example, dark energy acting as a kind of 'repulsive' gravitational force (see *What Is Dark Energy?*), there is nothing that proves exotic matter actually exists. Nevertheless, Alcubierre's work set people thinking.

Between 1996 and 2002, NASA funded a small research team called the Breakthrough Propulsion Physics group. It was their job to survey physics, looking for chinks in our understanding that might bring about a revolution in space propulsion. One of the things they looked at was the possibility of gravity control. For example, could we 'turn down' the gravitational pull of a launch pad so that a rocket could lift off more easily? Could we reduce the inertia of a spacecraft so that it can move more easily in space? These investigations led nowhere, but the team also looked at the concept of short cuts through space-time, intriguingly called 'wormholes'.

If you remember *Alice's Adventures in Wonderland*, you can think of a wormhole in space-time as the rabbit hole. To visualize how it works, imagine that a piece of paper is a two-dimensional Universe. To travel from one corner to the furthest corner, a two-dimensional inhabitant would have to walk the diagonal length of the paper. Now imagine that you fold the Universe around, so that the far corner, which the flatlander wants to reach, is sitting just at the corner where he starts. Now, all he has to do is hop upwards through the third dimension, which is a totally alien dimension to him, and he will miraculously find himself on the other side of the Universe. He has travelled through a wormhole, which is literally a short cut through a dimension that we cannot directly perceive.

Investigating the physics of wormholes, the physicists discovered that unfortunately, like Alcubierre's warp drive, they require exotic matter to make them traversable. The group concluded that no fundamental breakthrough in space travel was imminent, but they did identify anomalies in current knowledge that inspired new thoughts. Two of these anomalies, now well known to space engineers, hint at new forces and energy sources that have so far defied explanation.

Anomalous forces

The 'Pioneer anomaly' affects the Pioneer 10 and 11 space probes, which have been coasting away from us through space since their encounters with the giant planets Jupiter and Saturn in the 1970s. They are being mysteriously decelerated by a tiny amount, so every second their speed decreases by about one billionth of a metre per second. An international consortium of scientists and engineers are investigating the spacecraft's data in the hope of understanding exactly when the strange deceleration began, whether it built up gradually or just switched on, and whether it can be explained by some onboard malfunction. To date, they have explained most of the anomaly as heat from the radioactive generators onboard, but a minor part of the deceleration remains unaccountable and suggests that something fundamental in our understanding of gravity is not quite right.

The second unexplained effect is the 'fly-by anomaly'. More significant than the Pioneer anomaly, it has affected a number of different spacecraft, including the Jupiter-bound Galileo, the Near Earth Asteroid Rendezvous (NEAR), the Saturn probe Cassini and the comet chaser Rosetta. The anomaly appears in the form of an unexpected acceleration that imparts an extra few millimetres per second onto the velocity of a spacecraft when it flies past a planet. There have been attempts to explain this by errors in the mathematics, but the more spacecraft that experience the bizarre acceleration, the less likely it seems that so many different teams can all be getting their sums wrong. Other possible solutions include some kind of natural decrease in the spacecraft's inertia, making it more responsive to the gravitational field it is moving through.

It is not yet possible to say whether either of these effects will lead to a revolution in propulsion technologies, but understanding them is a prime focus for physicists. History has shown that profound scientific breakthroughs often begin with the recognition of small anomalies.

Time travel

If all the talk of warp drives and suchlike sounds fanciful, putting those ideas into practice will be a walk in the park compared to building a time machine. First, we need to understand what we mean by time. Time is an immensely difficult concept to define; unlike electrical charge or mass, it is not something that can be measured. This may not be obvious, as we are so used to monitoring the passage of time in our lives. But clocks do exactly this – mark the passage of time – by using some phenomenon in which time is an integral factor, such as the oscillation of a quartz crystal or the decay of a radioactive isotope; nothing actually measures time itself. Nor is time like shape, taste or colour, because it cannot be perceived by our traditional senses; yet we are constantly aware of its passage by the changing nature of events around us, or just by our own changing thought patterns. We are travelling through time, one-way, into the future.

The passage of time can be slowed down, however: special relativity tells us how to do this. As well as the increase in mass experienced when something travels close to the speed of light, another bizarre special relativistic correction involves time. Known as 'time dilation', this states that the faster an object moves, the slower time passes for it. This seems to offer something of a solution for interstellar travel because, if it were possible to accelerate a spacecraft to relativistic velocities, time would slow down inside it, allowing humans to reach the stars within their lifetimes. The downside is that outside the spacecraft time would continue to run at its normal rate and many years would pass, perhaps centuries. Imagine that one of a pair of identical twins becomes an astronaut and leaves Earth on a spaceship capable of travelling at a significant fraction of the speed of light. Upon his return, he has barely aged but his twin will now be an old man because he has experienced the passage of time differently.

General relativity, too, offers some ideas here. It describes how time is slowed down in the presence of a gravitational field. Where the field is weaker, for example at the altitude of the International Space Station, time would pass quicker: a clock on the space station would gain about 1 second every 10,000 years compared to an identical clock on the Earth's surface. This may sound small but with atomic clocks accurate to better than one part in a trillion, the time-dilating effects of general relativity can be easily be verified.

In science fiction, time travel has often involved travelling to the past in some sort of time machine. A far cry from the contraptions suggested by fiction writers, the time machine proposed by physicist Frank Tipler was a sound mathematical possibility. In 1974, he showed that a rotating cylinder would drag the space-time continuum around with it, rather like the way a spatula stirs up honey. If the cylinder were rotating fast enough, Tipler's calculations showed that routes into the past might be opened up. The trouble was that Tipler had to assume the cylinder was infinitely long. He suggested that a shorter cylinder might be capable of the same behaviour if it rotated faster. Interestingly, some of the recent theoretical analysis of rotating black holes mimics Tipler's mathematics, suggesting that time travel might be possible in the twisted space-time region close to a black hole known as the 'ergosphere' (see *What is a Black Hole?*). Other theoreticians, notably Stephen Hawking, have suggested that only exotic matter, with its negative mass, would be capable of opening routes into the past.

If time travel is possible then there are a number of paradoxes that immediately spring to mind, such as the scenario of going back in time and murdering a grandparent. A number of scholars have suggested that something would always happen to prevent logical paradoxes. But to most physicists this has the tinge of a supernatural hand of fate and they prefer to believe that either the Universe might split into parallel realities (see *Are There Alternative Universes?*) or that time travel is simply impossible.

Where is everybody?

Let us now return to that lunch with Enrico Fermi. After the brief discussion about the possibility of faster-than-light travel, the conversation turned to other topics and lunch continued. Fermi's mind continued to work though, and he suddenly exclaimed, 'Where is everybody?'. He elaborated by explaining that, if you assume that there are many extra-terrestrial civilizations in the Galaxy, then the extreme age of the Universe implies that some are older than human civilization and so should have developed space travel. As a result, the Galaxy should be teeming with technologically advanced civilizations and we should have been visited many times in the past and present. So where are they?

Fermi discounted the weird and wacky UFO sightings as evidence of anything, and his simple question became known as the 'Fermi Paradox'. He used it to argue that practical interstellar space travel was impossible; otherwise the evidence of extraterrestrial visitation should be all around us. The same argument can be advanced about time travel. If it is possible, someone somewhere at some time in the future will invent a time machine; at which point people will begin travelling into the past, and should be among us today. So, if the laws of physics really permit time travel, where are the visitors from the future?

The balance of evidence suggests that interstellar travel and time travel may indeed be impossible. But, before we become too pessimistic, think about technical knowledge and modes of travel just a few centuries ago. The natural philosophers of the 17th century no doubt believed that interplanetary travel was impossible, too.

Can the laws of physics change?
Physics beyond Einstein

The mathematics describing our world depends on universal constants of nature – but what if these are not constant? We may be forced to acknowledge that we as mere mortals are blind to the many dimensions of space-time that must exist beyond our perception.

Nearly two billion years ago, beneath what is now Gabon, Africa, in a region of the country called Oklo, an underground stream filtered through layers of sandstone rock. The water carried uranium and, over the course of many years, this radioactive element settled out into the sandstone and gradually built up into a seam of ore. At some point a geological upheaval thrust this vein onto its side and the flowing water began to erode the sandstone, concentrating the ore even more. Then, 1.7 billion years ago, the deeply buried uranium reached a critical mass and burst into life, becoming a natural nuclear reactor. It turned on and off repeatedly over the next few million years and eventually, as the mix of uranium isotopes changed, the reactor stopped.

In 1972, humans began to mine the ore to supply an increasingly uranium-hungry world and it was then that scientists first noticed some of the ore seemed to have already undergone nuclear fission reactions. As they analysed the ore and pieced together the evidence for Oklo having behaved as a natural reactor, another oddity surfaced. The nature of the nuclear reaction appeared to have changed and that could only happen if the laws of physics had changed too. Research conducted in 2004 showed that the strength of the force governing the nuclear reaction rate in Oklo had been different by a tiny amount, less than five parts in 100 million, from what it is today.

Ever since the time of Johannes Kepler in the early 17th century, physicists and astronomers have enjoyed unprecedented success in describing nature with mathematics. The equations derived have become our way of understanding the laws of physics and of predicting the behaviour of physical systems. Newton's Theory of Universal Gravitation, for example, was published in 1687. It tells us that the gravitational force between two objects is dependent on the two masses and the square of the distance between them. Is it possible that such a law could change over time? Perhaps one day it may depend upon the cube of the distance, or just half of one of the masses. This is surely a categorical impossibility. We can look out into the Universe and see celestial objects millions, even billions, of light years away that we can make sense of by applying the laws of physics as we understand them today. This strongly suggests that the laws apply throughout the whole Universe and that they do not change, or change very little, with time. Any possible changes must be very subtle: they cannot be changes in the mathematics, but the eyes of suspicion fall on the 'constants'.

The constants of nature

There are many so-called constants in nature. They are the values that cannot be derived from theory, and so can only be determined by measurement. They are used in the laws of physics as conversion factors to create exact mathematical relationships between quantities. In the case of gravity, mass in kilograms and distance in metres are equated to a force in newtons by Newton's 'gravitational constant'. This is often referred to as 'Big G' because it is denoted in the equation by a capital G.

Some of the constants are self explanatory, such as the speed of light. Others seem more abstruse, such as the Planck constant, which governs the way nature breaks energy up into small 'packets'. Despite calling these quantities constants, there has been a creeping suspicion over the last 15 years or so that some of them might be changing slowly with time – particularly the speed of light.

In 1993, physicist John Moffat published his solution to the cosmological horizon problem. This is the tricky observation that the temperature of the cosmic microwave background is virtually the same regardless of which direction we look (see *How Did the Universe Form?*), so completely disconnected regions of the Universe have somehow reached the same temperature. Traditional physics can only explain this if the Universe was driven into a sudden period of exponential expansion, known as 'inflation'. However, inflation does not have any firm foundations in physics, meaning that what actually drove this supposed expansion remains a mystery. This shortcoming led some to look for alternative theories of temperature equalization. Moffat pointed out that if the speed of light had been higher in the past, photons of light could have travelled much further and so could have equalized the temperature across a much wider expanse of space without the need to invoke inflation.

Other physicists used the same idea to perform a new analysis of the cosmological flatness problem (see *How Did the Universe Form?*), and showed that they could also account for this without inflation, providing that the speed of light was extremely high during the first moments of the Universe's life and then fell quickly to near its modern value. Astronomers cannot directly test for such a circumstance since they cannot see the fleeting period of time immediately after the Big Bang. But they can study distant quasars – early galaxies powered by matter falling into black holes (see *What is a Black Hole?*) – in the hopes of catching the last vestiges of any change in the speed of light. In order to detect such a change they look at something called the 'fine-structure constant', which defines the strength of the electromagnetic force in relation to the other forces of nature and determines the pattern of spectral lines from light sources (see *What Are Stars Made of?*). Its value depends on the speed of light and, importantly, it is what scientists call 'dimensionless'.

Dimensionless constants

Physicists have to be careful when drawing conclusions from the measurement of constants that have units attached to them. For example, the speed of light is measured in metres per second, or any other units of length and time chosen. If a variation is measured, the researcher cannot be sure whether it is the speed of light that has truly varied; it could just as well be the rate at which the clock has ticked, or the length of the ruler, that has changed. To avoid this confusion, physicists tend to concentrate on examining dimensionless constants when searching for natural variability. If you measure the ratio of, say, the proton's mass to the electron's mass, then the units – kilograms – will cancel out and the resulting constant you get will simply be a number. If something weird happens to the way you define a kilogram, that will be cancelled out in the ratio and not affect your conclusion. So, if the value of the ratio changes by even the smallest amount, you can be certain that at least one of those masses is actually changing in some way.

The fine-structure constant is just such a dimensionless constant. It is obtained by combining the speed of light with the Planck constant of energy and the charge on an electron. It affects the outer structure of each atom, which controls the way the atomic electrons react with passing light beams. If the speed of light were to change with the passage of time, the fine-structure constant would also change, and the characteristic spectral lines of all the atoms would change as well.

This is exactly what one group of astronomers believe they have seen. In 1999, John Webb of the University of New South Wales used the world's largest optical telescope to observe 128 quasars out to distances of around 10 billion light years. Webb's team collected the quasar light, split it into spectra, and looked for the fingerprints of intervening atoms. Their analysis showed that the spectral lines changed in a way that was consistent with the fine-structure constant having increased slightly during the course of cosmic history, by around 1 part in 100,000 during those 10 billion years.

Numerous groups are now working to verify, or disprove, the variation of the fine-structure constant. If they confirm it, then scientists will have to decide which of the constants that define it is actually varying. Is it the speed of light, the charge on an electron or the Planck constant of energy? Most people suspect it is the speed of light, because of the way a change could help solve the horizon problem of uniform temperature across the Universe. Nevertheless, the discovery of a constant changing has enormous consequences for our understanding of the Universe. It points to physics beyond Einstein, perhaps even to the elusive 'theory of everything'.

Most physicists believe that the best candidate for a theory of everything is string theory (see *Was Einstein Right?*). This complex mathematical theory replaces particles with wiggling strings, but the wiggling take place in higher dimensions than the three we are directly familiar with. We see the strings as particles because, rather like icebergs, there is a lot going on 'beneath the surface'. According to string theory, only if all the higher dimensions are taken into account will the value of physical constants remain truly constant. Hence, string theory allows for the constants of nature to appear to change in the dimensions that we perceive. If we could measure a change, it would allow us to use string theory to prove the existence of higher dimensions and to see how they are behaving.

Big G

The strength of gravity has been another target for physicists searching for variations in the constants of nature. The difficulty is that Big G, which encapsulates the strength of gravity, is one of the most elusive constants in nature because despite sculpting the Universe on its largest scales, gravity is the weakest of the forces. This means that Big G is difficult to measure accurately. It was over a century after Newton's derivation of universal gravity that the first successful measurement of Big G was made, and another century before the realization sank in that this is what had been achieved.

Working in 1797, Henry Cavendish managed to measure the force of gravity generated between two lead balls, one 12 inches in diameter and the other much smaller. He did this using a piece of equipment called a torsion balance, which transformed the small gravitational force between the two weights into a twisting of the apparatus, which could be seen and measured. Cavendish then weighed the smaller lead ball, which gave him the force of the Earth's gravitational field, and compared the two forces to obtain the density of the Earth. This was a quantity much desired by astronomers of the time, because they could use it to calculate the density of the Solar System's other objects.

It was only in the late 19th century that scientists came to view Newton's gravitational constant as something fundamental to science. And then they went back to Cavendish's data to calculate a value for Big G. Since then Big G has been measured to greater and greater accuracy, even though there have been some hiccups along the way. In 1987, scientists thought Big G was known to an accuracy of 0.013 percent. Improved experiments in 1998 forced this to be re-assessed to a lesser accuracy of just 0.15 percent. Even today, the value of Big G is extraordinarily imprecise when compared with the force of electromagnetism, which is known to 2.5 million times greater accuracy than is gravity. It is this lack of precision that has led to most of the speculation about whether the constant might be changing in value slowly over time, in effect changing the strength of gravity. Such a variation would gradually change the orbits of stars and planets, affect the sizes of celestial objects, and even determine how brightly stars shine.

Most recently, lunar laser ranging experiments (see *Was Einstein Right?*) have shown that the value of Big G cannot have changed by more than one part in a million per year, otherwise their sensitive measurements would have picked up this variation in their 40-year observation of the Moon's orbit. This does not mean that Big G has remained constant; it simply indicates that any variation has been smaller than one part in a million. So astronomers continue to collect lunar laser ranging data in an attempt to extract ever-finer measurements and search for long-term effects. At the same time, other physicists are searching for temporary changes in the strength of gravity brought on by the movement of Earth around its orbit. This could give us a clue to the physics beyond Einstein.

Beyond Einstein

Einstein's theories of relativity rest upon the central tenet that the laws of physics are the same, no matter where or when you are located in the Universe or how you happen to be moving. How to transform what one observer can see into the viewpoint of another is known as the 'Lorentz transformation'. It is named after Hendrik Antoon Lorentz, who caught a glimpse of this behaviour earlier than Einstein. He derived a mathematical expression to describe it, but did not know how to correctly interpret it. If the constants of nature change, the Lorentz transformation no longer works precisely, and a 'Lorentz violation' is said to have taken place.

The great success of Einstein's laws in predicting behaviour in the Universe tells us that any Lorentz violation must be small. (This is no surprise: the tiny variation in Mercury's orbit, which showed where Newton's theory diverged from reality, was so small it went unnoticed for about a century and a half after Newton's theory was published.) String theory, however, allows small Lorentz violations to have taken place in the Big Bang. If they did, these will have imprinted themselves on the fabric of space-time and could make the laws of physics vary not over billions of years, but at minuscule levels, for example over the course of a single year as the Earth orbits the Sun and so travels in different directions through space. Think of this as a bit like the Earth orbiting on a hill, the gradient making it a little harder for the Earth to go uphill than to roll downhill. This would show up in the speed at which a ball falls to the ground: it may take slightly longer when the Earth is travelling 'uphill', than when it is coming 'downhill' six months later. Effectively, as the Earth travels in different directions through space in the course of its orbit, Earth's gravity will change by a minuscule amount. The obvious way to test for this is to drop objects throughout the year and measure their rate of fall. Comparing measurements taken six months apart should yield the greatest difference because then the Earth is travelling in opposite directions. The best place to conduct the experiment is in space, because when an object is in free fall, small gravitational variations can be measured very precisely. A number of missions hoping to pursue this research are currently on the drawing board.

Physicists will continue to search for changes in the constants of nature – both long-term and short-term effects – for as long as they believe that string theory is the way to unite gravity with the other forces. By measuring the amount of change, they will be able to home in on the correct version of string theory and understand better its picture of a multi-dimensional Universe. Newton's theory of gravity is said to have been inspired by watching an apple fall to the ground; earlier, Galileo is said to have dropped objects from tall buildings to discover that all objects fall at the same speed, regardless of their composition or mass. It may not be coincidental that our next breakthrough in understanding the Universe could come from measuring falling objects in orbit.

Are there alternative universes?
Schrödinger's cat and the implications for us all

There is an exact 50-50 chance that the cat is alive, or is dead. Until we look at the cat, we do not know. Can we know? Can the state of the cat be described before we look at it, other than by a quantum superposition of 'alive' and 'dead'?

It must rank as one of the most famous cases of animal abuse in history. Thankfully, though, it is an entirely fictional thought experiment. Called 'Schrödinger's cat', the idea is that a cat is put in a sealed box with a device containing a radioactive atom that has a 50-50 chance of decaying within an hour. If the atom decays, a flask of poison is broken and the cat dies. The Austrian physicist Erwin Schrödinger constructed this macabre 'experiment' in 1935 to make a point: he thought that although the newly developed quantum theory could predict the behaviour of particles, it could not be a true description of reality because it led to bizarre consequences. Specifically, quantum theory allowed particles to possess contradictory properties until they were measured. To highlight what he saw as an insoluble problem he asked how the contents of the box could be described at the end of the hour before somebody peeped inside.

The half-alive, half-dead cat

The condition of the cat obviously depends upon whether the radioactive atom has decayed or not, and this decay depends on probability. The physics of probable states is the basis of much of quantum theory (see *What is a Black Hole?*). In Newtonian physics, everyday objects interacted in ways that were rigidly repeatable and as predictable as clockwork. When someone threw a ball in the air, there was no doubt that it would fall back to the ground; the only question was how long it would remain in the air and Newton showed that this could be precisely calculated. On the scale of individual atoms, however, probability enters the mix and physicists have to use quantum theory. This shows us that events are not rigidly repeatable in the subatomic realm, so probabilities need to be ascribed to outcomes. Hence, in radioactive decay, there is a calculable chance that an atomic nucleus will decay within a given time, but there is not a guarantee. What makes the nucleus 'decide' to decay at a particular instant is completely unknown to us; we simply have to accept that probability is hard-wired into the Universe. Anything that relies on the inherent uncertainty of subatomic processes is said to be a quantum system. Returning to Schrödinger's cat, quantum theory can provide an equation that perfectly describes the radioactive atom in a state that represents both decayed and yet-to-decay possibilities. But what does this mean for the cat? Must the cat be thought of as simultaneously alive and dead until the box is opened? Schrödinger considered this ludicrous.

Once the box is opened, the mystery is solved. The cat will be either alive or dead, depending on whether the radioactive atom has decayed or not. But this leaves us with another dilemma: what happens to the 'unused' state of the atom, the alternative possibility that was not observed? Did it simply cease to exist when the box was opened?

Danish physicist Niels Bohr did indeed think the alternative state just vanished. Puzzled by how to interpret the mathematics of quantum theory, he and colleague Werner Heisenberg had decided in 1927 that the act of observation forces the quantum system to 'make up its mind', and become one thing or the other. Prior to the observation, the quantum system was in a mixed state, a 'superposition' as physicists call it, of all possible outcomes. This is known today as the 'Copenhagen interpretation'.

At its heart, the Copenhagen interpretation says something exceptionally profound – that measurement creates reality. The common-sense problem with this view is, as

Schrödinger pointed out, that we have a zombie cat for an hour, half-alive half-dead, until the box is opened and somehow the act of observation makes it either dead or alive. Instinctively this sounds wrong – why should the act of observation be essential in creating reality? Surely we have to believe that the Universe is 'solid', even when we turn our backs on it. For the first time, physicists were faced with a fundamental problem that verged on philosophy: does quantum theory describe reality or is it just a mathematical trick that gives the right answer?

The 'many-worlds' interpretation

Trying to square the mathematics with reality has perplexed physicists for the better part of a century. In 1957, American physicist Hugh Everett proposed that we should take quantum theory at face value and believe that its mathematics does describe reality. Therefore, when the equations show different possible outcomes, all of them must be played out somewhere. Everett had no idea where these alternative realities might be located, but in one 'universe' the cat would live; in another it would die. Our own Universe meanders from one quantum decision to the next, tracing just one of a multitude of paths through reality.

Everett's idea has become known and respected as the 'many-worlds' interpretation, but it was slow to catch on, mainly because Bohr refused to take it seriously. It was in the 1970s that physicists became interested in the hypothesis, because they were starting to use higher dimensions in their calculations, and they realized that these might provide locations for parallel universes, coincident with our own but shifted through a dimension we cannot perceive. Perhaps, they wondered, it was in these parallel universes that Everett's 'many worlds' could exist. It was the equivalent of letting the genie out of the bottle. Physics is now alive with the idea of parallel universes; there is even evidence pointing to a variety of different types. This panoply is known as the 'multiverse'.

An infinite Universe with infinite possibilities

In 2003, American physicist Max Tegmark classified the possible alternative universes into four different types. The first and simplest case comes about because our Universe may be infinite in size. Measuring the size of the Universe has been a preoccupation for astronomers since the 16th century. Every time they have devised a method to measure more distant celestial objects, they have been astounded by just how far away they are. In other words, the Universe has continually surprised us with how big it is, and this has led to the suspicion that it is in fact infinite. We cannot possibly see all of it, because, in the 13.7 billion years since the Big Bang, light can only have arrived from regions that were once no further than 13.7 billion light years away. Anything beyond this is impossible for us to observe at present.

Astronomers have been able to test whether the Universe is *smaller* than 13.7 billion light years, by looking for repeating patterns in the ripples of the cosmic microwave background radiation. To see why this would reveal a 'small' Universe, think of the Earth's surface. Forget for a moment that we perceive three dimensions, just take it for granted that the ground is flat. Start walking north; the ground continues to appear flat around you, and you keep walking. Ignore the obvious hindrance of the polar ice caps, oceans and mountains; just keep walking. Eventually, you will circle the Earth and arrive back where you started. Such a shape, in this case a sphere, is known as an unbounded but finite shape because you do not come across any boundaries but the surface of the Earth is finite in area. The Universe might be similar, the key being that space may be curved through a higher dimension than we can perceive. In our example of the Earth's surface, it was curved through the third dimension, which we were pretending we could not sense. In the case of the Universe, it is curved through the fourth dimension (which Einstein introduced to explain gravity).

If the Universe curved completely back on itself in the fourth dimension, then a powerful telescope on Earth could, in principle, see the Milky Way Galaxy apparently very far off in space. But of course the light would have been travelling for billions of years and so the Milky Way would appear much younger than it is today. It would be like seeing all the way round the curve of the Earth and observing the back of your head far off in the distance – but you yourself would be a baby.

Looking for repeating patterns in the cosmic microwave background radiation is the somewhat more practical equivalent of searching the distant galaxies for a young Milky Way. To date, no such repeats have been found and this is taken as evidence that the Universe extends further than we can see – beyond 13.7 billion light years. The theory of inflation, that the Universe underwent a sudden period of exponential expansion just after the Big Bang (see *How Did the Universe Form?*), if proven, would mean that the Universe must be vastly bigger than this; in fact most cosmologists believe that inflation leads to an infinite Universe. Even if inflation is proved wrong, an infinite Universe is still a possibility.

If the Universe *is* infinite then every outcome – no matter how unlikely – is played out somewhere. Somewhere in the Universe there is an alternative Earth where the alternative 'you' wrote this book and the alternative 'me' is reading it. Think of any possibility that does not contravene the laws of physics and it will have happened. Tegmark calls these level I parallel universes. They possess the same laws of physics but started from different initial conditions, hence they are not quite the same. As time goes by, light will arrive from further and further away and these remote regions will come into view, gradually revealing these alternative universes to us.

Chaotic possibilities

A slightly different version of inflation, known as 'chaotic inflation', makes it possible that new universes sprouted away from our own because of the way that quantum theory works. If this happened, it would have set into motion a chain reaction that continues somewhere in the multiverse today – in short, a never-ending sequence of other universes being born. These comprise Tegmark's level II parallel universes. Unlike the level I alternative universes, they do not just lie a long way away, but inhabit entirely different dimensions of space.

In one of these level II parallel universes, the way the forces of nature evolved may be different from the way this happened in our Universe, and so the strength of these forces could be different. This would be reflected in the constants of nature taking on different values from the ones in our Universe. Perhaps gravity is a bit stronger and so stars are pulled together more quickly, burn more brightly and hence live shorter lives. Or the strong nuclear force is a little weaker, making more atoms radioactive; this would generate more heat inside planets, creating more volcanic activity.

Some of the parallel universes may be 'flat', with only two extended dimensions, whereas some may have four spatial directions, or six, or none. There is no fundamental difference between the other parallel universes and our own; all have the same laws of physics. We believe that a universe transforms itself, from the high-energy state just after the Big Bang to the low-energy state of today, in an essentially random process. The strengths of the forces, the number of dimensions, even the variety of particles, are all fixed by this unpredictable process. Think of it as a ball whizzing around the top of a roulette wheel. Each ball that the croupier spins begins in a high-energy state indistinguishable from any other spin of the wheel. When it loses energy and falls down into the wheel it eventually ends up in one of the numbered slots, each one an equally valid outcome. In the case of a roulette wheel there are 37 or 38 different pockets, but for a collapsing universe there are an infinite number of possibilities. So some level II universes will be similar, even identical to ours, whilst others will be vastly different.

Weirder parallel universes

In his next category, the level III parallel universes, Tegmark turned his attention to Schrödinger's cat and Everett's original suggestion of 'alternative realities' in which every possibility is played out. Tegmark found a subtle but important difference between these parallel universes and the previous two types. It is to do with how they are created.

According to Everett's many-worlds interpretation of quantum theory, the Universe splits when a quantum decision is revealed – such as when the box in Schrödinger's cat experiment is opened. This whittling of possibilities into a certainty is known as 'decoherence' and, until the mid-1990s, physicists did not know how it happened. In refuting Everett's ideas, Bohr had talked cryptically about the act of observation being needed to force the quantum system to make its decision, but he failed to say what defined an observer. Was the cat in Schrödinger's thought experiment an observer of its own condition, or was human self-awareness required? Could particles themselves be 'observers' by dint of their physical interactions?

An experiment conducted by Serge Haroche and colleagues in 1996 using rubidium atoms and microwaves provided the answer by showing that decoherence occurs as atoms interact with their surroundings. No intelligent observers were needed, just the random interaction of other particles. The conclusion therefore is that particles are indeed Bohr's 'observers' and that the act of observation is equivalent to a physical interaction between particles. This solves the worst aspect of Schrödinger's cat experiment, namely the period in which the animal is half-dead, half-alive. This scenario never happens because the interaction of the particles inside the box – the atoms in the cat, the radioactive particle, the molecules in the air and in the poison – will mean that if the bottle breaks and releases the poison, the cat will be killed instantaneously as our common sense would suggest.

The many-worlds interpretation can be thought of as offering a 'life after death' for the timelines that are rejected in our Universe. As one possibility comes to an end for us, so a new universe springs into existence to play it out. But there is a remarkable coincidence here: the possibilities played out in the level III parallel universes will be identical to those played out in level I examples. It is just that level I universes are separated from us by vast tracts of space. Level III universes supposedly pop into existence, presumably in some other dimension, as our reality unfolds. As yet, no one can reconcile these two similar yet different concepts.

It might be thought on reading this that every time we make a conscious decision other universes are conjured up, but such an extrapolation would be wrong. This behaviour is restricted to quantum processes. The only way our conscious decisions could create alternative realities is if somewhere deep inside the brain a decision is based upon a single quantum particle that spontaneously changes its state. This tiny happening would then need to be amplified in our consciousness to become a decision. Intuitively, this feels wrong, since decisions seem like a much higher processing of information. Each of us makes decisions by weighing evidence and past experiences and 'computing' what we hope will be the best course of action, not by a random alteration of a particle from one quantum state to another.

Flipping a coin to provide an outcome is not a quantum process either. A coin flip is governed by factors that are hard to predict, but not because of quantum uncertainty. It takes place on a scale much larger than the scales at which quantum effects apply. So, sadly, deciding whether or not to finish this chapter is not going to cause an alternative universe to spring into reality – you may as well read on.

Why these laws of physics?

Suppose that physicists succeed in finding a theory of everything that describes the 'superforce' that controls the Universe and gave rise to the physics of today. We would be forgiven for thinking that their job was done. Far from it – the hard work could just be starting because they would then have to set about answering the really important question: 'Why should the laws of physics be as they are?'. The final level of parallel universes rests on the answer to this question and is perhaps the most philosophically influenced.

It was originally hoped that string theory (see *Was Einstein Right?*) would answer the question and tell us why the Universe behaves the way it does; that it would provide some deep reason why the constants have the values that they have and the laws are as they are. That hope faded because there appeared to be many different versions of string theory, all of which seem equally plausible. By 1995, physicists had developed between them five distinct string theories, each one using ten dimensions in its calculations. In trying to decide which of these competing theories was correct, they found something remarkable. The various ten-dimensional string theories could be shown to be different manifestations of the same, much broader theory, if they simply added another dimension to the mathematics. They called this eleven-dimensional model 'M-theory', perhaps for Mother theory although no one seems to recall the reason; some have even suggested Magical.

Physicists now envisage the many possible string theories that can come out of M-theory as a landscape, where valleys contain self-consistent universes and mountains represent energy barriers between them. Our Universe sits in one of the valleys, but as yet we do not know which string theory describes it, or indeed whether string theory is correct at all. The other valleys in the M-theory landscape all have different laws of physics and different constants of nature, and are the level IV universes. Some of them will be similar to our Universe, others will be vastly different, and many physicists now suggest that every combination of laws is tried out in one of the parallel universes. Some scientists even argue that if you reject this interpretation of the multiverse then your only recourse is to believe that God created our Universe (see *Is There Cosmological Evidence for God?*).

Proof of parallel universes

A strategy for how to search for level III and level IV universes currently eludes physicists, but there is a way to verify that level I and level II universes exist. Astronomers are currently searching for proof that inflation happened. If they find it, they will also have proof that parallel universes exist, because any form of inflation is thought to create level I parallel universes, which would be very far away in space and have the same values for their physical constants. If the variation known as chaotic inflation is confirmed, there will also be level II universes, 'budding' off from our Universe and having different values of the physical constants.

Inflation leaves its mark on the cosmic microwave background radiation in the form of fluctuations in the density of the cosmic matter and in the orientation of the cosmic microwave background rays. Many observations of the microwave background have been broadly consistent with inflation, although some discrepancies have been uncovered. The recently launched European spacecraft *Planck* will investigate further by taking highly accurate pictures of the microwave background. The orientation of the microwave radiation is known as 'polarization' and should have been imprinted on the microwaves during inflation by gravitational waves coursing through the Universe in the split second after the Big Bang. The gravitational waves moved through the fabric of space-time like ripples on a pond; as they passed, they would have alternately squeezed and then stretched matter. The different versions of inflation theory impose different patterns on the microwave background.

In March 2014, astronomers running the BICEP2 experiment at the South pole announced the discovery of these patterns. Although the result must now be verified by other experiments and Spacecraft looking for the same pattern, if inflation of one type or another is proven to be true, then physicists will have to accept the fact that parallel universes do exist. And all of us will have to come to terms with the idea that there are many different versions of each of us out there somewhere.

What will be the fate of the Universe?

Big crunch, slow heat death or big rip

The next time you are eating a piece of Swiss cheese, think about the fate of the Universe. The cheese may just hold a clue as to how the Universe will end; not the cheese itself exactly, but the characteristic pattern of holes within it. Matter in the Universe appears to be distributed in a rather similar way – there are titanic voids that contain hardly anything, surrounded by large tracts of dust and gas where the galaxies form. Understanding this distribution is crucial to determining the Universe's ultimate fate.

Almost as soon as astronomers came to accept in the late 1920s that the expansion of the Universe was a reality, it led them backwards to a creation moment, the now familiar 'Big Bang'. With that in mind, they began looking forwards and wondering how the Universe might end. The basic mechanics of the problem seemed straightforward: the Big Bang sent everything bursting outwards but the gravity of the celestial objects would attempt to pull it all back together again. So what might happen seemed to rest on how much gravity-generating matter there is in the Universe, and how it is distributed. Two scenarios are possible: if the overall density of matter is greater than a certain critical value then gravity will overcome the expansion and the Universe will contract, sending all celestial objects crashing back into each other; if the density is lower than the critical amount then the Universe will expand for the rest of time, its galaxies becoming ever farther apart. Astronomers refer to the first scenario as the 'closed' Universe and the second as the 'open' Universe.

The big crunch

General relativity tells us that the density of matter governs the overall curvature of space (see *How Did the Universe Form?*). Imagine for a moment that the Universe has just two dimensions: length and breadth. The closed Universe scenario can be represented by the surface of a ball and is mathematically described as negatively curved. This is the kind of Universe known as unbounded but finite; you could travel all the way round it eventually arriving back at your starting point (see *Are There Alternative Universes?*). Such a Universe could be extraordinarily big but not infinite in size. If it were infinite it would not curve back on itself – it would extend in all directions forever.

So let us assume we live in a closed Universe. Since it cannot expand to infinite size, at some point in the future it will stop expanding and begin to contract. At this point the redshift of distant galaxies (see *How Big is the Universe?*), caused by the expansion of space, will reverse. As the galaxies begin to fall back toward us, their light will be squashed to shorter and shorter wavelengths, making the stars in those galaxies appear to blaze with blue brilliance, as if they were shining at a much higher temperature than normal. Astronomers term this phenomenon 'blueshift'; eventually, as the galaxies fall towards us faster and faster, the blueshift will become more and more pronounced, squashing the starlight into ultraviolet rays and then X-rays.

Galaxies will be pulled closer and closer together, heading for the 'big crunch'. A billion years or so before this super-collision, clusters of galaxies will merge together. Around 100 million years before the end individual galaxies will begin merging. For the last million years of the Universe's existence, there will be no such thing as an individual galaxy - the entire Universe will be one great ocean of stars. By this time, the blueshift will be having a highly noticeable effect on the cosmic microwave background radiation; it will transform the microwaves first into infrared and then into visible light, making the whole night sky light up around 100,000 years before the now inevitable big crunch. Finally, the blueshifted background radiation will become so intense that it will exceed the temperature of the stars themselves. The stars will dissolve into space and the Universe will resemble the fireball of the Big Bang.

These spectacular death throes are, in some ways, a rewind of the Big Bang. Some cosmologists have speculated that something may eventually prevent the Universe from collapsing completely and make it rebound, turning the big crunch into another Big Bang and starting the whole process of cosmic evolution off all over again. If so, it would be like rebooting the Universe.

Slow death

The alternative scenario is that of the open Universe, where the density of matter is less than the critical value. The shape of space in such a Universe is more complex; the best two-dimensional analogy is rather like the upper surface of a saddle, with one dimension curving upwards and the other curving downwards. Unlike a saddle, however, it extends infinitely in all directions. The open Universe will expand forever, but this does not mean that it will remain more or less the same forever.

Stars will continue to live and die much as they have done over the last 13 billion years, but space itself will change around them. As space continues to expand, galaxy clusters will be driven ever farther away from each other. Most of the galaxies that draw our attention today will eventually be lost from sight; only the stars in the Milky Way and in the 30 or so nearest galaxies in our own cluster will remain visible to us. For all the rest, their light will be redshifted away from visible light, into the infrared and then into the weak radio region of the spectrum. Perhaps the only clue future beings will have of the existence of other galaxies will be a faint radio hiss coming at them from all around, similar to the cosmic microwave background of today. The already weak background radiation itself will have been redshifted too, rendering it unobservably feeble. Future civilizations will thus have no evidence that the Big Bang ever took place (see *How Did the Universe Form?*).

Within each galaxy cluster, where gravity can resist the cosmic expansion of space, the individual galaxies will all eventually coalesce. As the galaxies collide, some stars will be flung out into intergalactic space to wander the Universe alone. Others will be catapulted into the supermassive black hole at the centre of their galaxy, temporarily reigniting the black hole and making the galaxy active again with a blazing core. This could well be the ultimate fate of our own Sun, as the Milky Way Galaxy and the Andromeda Galaxy collide. Every galaxy, it is believed, has a supermassive black hole and as these vortexes of space-time interact, releasing a blaze of cosmic energy, an even more massive, voracious cosmic dustbin will be forged. After perhaps 100,000 billion years, all the cosmic gas will have been either pulled into existing stars or sucked into gigantic black holes.

With no gas clouds left to collapse into new stars, stellar activity will be coming to an end. The lights will be going out across the cosmos, leaving space to contain nothing but highly isolated collections of dead stars: white dwarfs, neutron stars, and black holes both large and small. These stellar cadavers will occasionally collide, releasing a sudden flash of brilliance, but, by and large, no light will be shining through the Universe.

But this might not be quite the end. There are hints that protons, which make up a vital constituent of each atomic nucleus, may not be permanently stable. If they eventually decay, all the atoms in the Universe will disintegrate, leaving a sea of subatomic particles. Chemical reactions and nuclear reactions will become impossible, and the only structures able to form will be black holes as the particles clump together.

Even the black holes may not last forever. They may gradually evaporate, radiating their matter content back into the Universe in the form of more subatomic particles. Such a phenomenon, known as 'Hawking radiation' after its proponent (see *What is a Black Hole?*), would take place over an inconceivably long time period, perhaps a googol years (1 followed by 100 zeros). So, the long-term fate of an open Universe is to become a dilute sea of particles, all at approximately the same low temperature and unable to react with one another. Such a state is known as the 'heat death' of the Universe.

The big rip

We have talked of two scenarios, but between the open Universe and the closed Universe eventualities is the unique possibility of a 'flat' Universe. Its two-dimensional analogy is simply a flat sheet that extends forever in all directions. Whilst there are myriad types of both open and closed Universes, depending upon the density of matter, there is only one flat Universe. It corresponds to the Universe containing exactly the critical density of matter. Its ultimate fate would be identical to the open Universe scenario – the dilute sea of cold particles – but it is a highly improbable case, given how finely tuned the Universe would need to have been during the Big Bang to create such a precise amount of matter. Nevertheless, it is the type of Universe that cosmologists believe that we live in. Astronomers have been able to analyse ripples in the microwave background to discover the geometry of the Universe, and have found it to be almost perfectly flat. But this result did not balance with their inventories of matter until they discovered the acceleration of space and decided that some form of 'dark energy' must also permeate the cosmos, making up the deficit (see *What is Dark Energy?*).

Until we know exactly what dark energy is, we have no way of being certain how it behaves and so the straightforward picture of a flat Universe expanding forever may not be true after all. Dark energy may turn off, gain strength or even reverse its effects. Astronomers need to be able to determine which behaviour pattern dark energy will follow; until then they cannot rule out any possibility for the fate of the Universe. If the dark energy effect remains constant, the expansion of space will continue to accelerate, shortening the time it will take for the galaxies to disappear from view. But if the dark energy were to reverse its behaviour, it would increase the strength of gravity in the Universe and could pull everything back into a big crunch.

The most bizarre option would be if dark energy continually increased with time. If this were to happen, the Universe's expansion would accelerate at an ever-increasing rate. After driving all of the galaxies so far away that we could no longer see them, dark energy would then go to work on the Milky Way. It would disrupt our Galaxy and even drive planets out of their orbits around stars. Ever-strengthening dark energy would then pull apart the stars, followed closely by the planets. Finally, it would explode the very particles that make up matter. Astronomers call this nightmare scenario 'the big rip'.

Sudden death

This may all sound bleak, but string theory offers a number of even more bizarre scenarios – at least one of which, called 'vacuum decay', could result in the sudden death of the Universe at any time. If string theory is correct, there could well be a multitude of Universes (see *Are There Alternative Universes?*), each one separated from the others by energy barriers. According to quantum theory, it may be possible for a tiny region of our Universe to 'tunnel' through to another Universe with a lower energy state. If this were to happen, our whole Universe would transform itself into this lower energy state (rather like a glass of water can turn itself into a glass of ice if the temperature drops low enough). The entire fabric of space would undergo the transformation, and this would destroy everything. The catastrophe would spill out from the epicentre, in all directions, at the speed of light. This could give us billions of light years of notice if it begins far away, but there will be nothing that we can do to stop it. Our fate will have been sealed.

Another idea, also relying on the 'multiverse' proposed by string theory, emerges from attempts to understand what created the Big Bang. It envisages a collision between a parallel Universe and our own, which would excite both Universes into high-energy states. The theory, known as the 'ekpyrotic' Universe after the Greek word meaning 'out of fire', suggests that the two Universes are joined by a piece of cosmic elastic and continually bounce into each other, like a pair of clapping hands. When they touch, everything is erased and the Universes are then reborn in a new Big Bang. The theory does away with the need for inflation after the Big Bang, and explains the ripples in the microwave background as the slight differences in time at which various parts of the two Universes collided. There would be no warning of when the two Universes would collide. One moment all would seem normal, the next: oblivion. But it may be comforting to know that a new Universe would be created from the ashes of our own.

It seems somewhat depressing that there are currently no hypotheses that preserve the Universe in its present state. The 'steady state theory', popular in the mid-20th century, relied on the continuous creation of matter to fill the gaps created by the expanding Universe. But the theory was disproved by the discovery of the cosmic microwave background radiation and its interpretation as the residual energy of the Big Bang. The majority of astronomers now believe that the most likely fate for the Universe is to expand forever and suffer a heat death. But none of the scenarios presented in this chapter can yet be ruled out.

Is there cosmological evidence for God?

The apparent fine-tuning of the Universe for human life

In the course of his confrontation with the Vatican authorities, Galileo Galilei made a flippant comment that nevertheless defined the line between astronomy and religion. He said, 'The Bible tells us how to go to Heaven, not how Heaven goes.' Since Galileo's time, astronomy has frequently clashed with religion over whether God exists. Some consider that the laws of our Universe are improbably finely tuned to allow the creation of life. But if our Universe is just one facet of the multiverse, then there is nothing special about our Universe and there seems no need for God.

Galileo meant by his remark that the Vatican theologians should not attempt to determine the workings of the cosmos by interpreting passages of the Bible. Only astronomy and rational thinking could reveal the way the Universe behaved, he tried to tell them, but this need not detract from the Bible's role in leading humankind to the salvation of their fragile souls.

When Isaac Newton published his Theory of Universal Gravitation in 1687, he too was attacked on religious grounds, for promoting atheism, because his theory appeared to explain all motion on Earth and in the Universe without the need for God. A religious man himself, Newton stated in response that his theory did not explain gravity, but merely described what it did, and that perhaps God was to be found in the explanation of gravity's nature. Centuries later, Albert Einstein said that he could not believe in an interventionist deity, one that is consciously manipulating reality, but he could believe that the laws of physics were an embodiment of God.

These thoughts of great scientists highlight the difference between science and religion: science proceeds from a basis of unambiguous definition that is testable and repeatable, whereas religion is more malleable in its founding princi- ples. In deciding whether there is a God or not, a scientist's first question would most likely be 'What do we mean by God?' In other words, how do we define God?

The meaning of God

The Bible talks of miracles, such as the parting of the Red Sea and turning water into wine. Taken literally, these stories strongly suggest that God sits outside the laws of physics. In this case, strictly speaking, no scientific argument can be used to discuss God because He is able to subvert nature to His will: He is supernatural. The argument that today's scientists would raise against such a supernatural deity is that they see nothing in the Universe that requires such a presence.

Early scientists, however, would talk about investigating nature in order to find God's place in the Universe. They hoped to uncover phenomena that could not be explained rationally and that therefore required God's supernatural intervention. As science became more powerful, however, it could explain more. For example, in the 16th century there were those who wondered why the planets move as they do; most thought it was the will of God. Resolving the question was impossible until Tycho Brahe spent his life recording the nightly position of the planets, so that Kepler could derive three laws that encapsulated the apparently disparate planetary motion (see *Why Do the Planets Stay in Orbit?*). Subsequently, Newton explained that Kepler's laws of planetary movement were the result of a force called gravity; then in the 20th century, Einstein explained what gravity was and that there must be a fabric of space expanding in all directions (see *Was Einstein Right?*). Astronomers realized that this implied that the Universe began in a moment of creation, now known as the Big Bang (see *How Did the Universe Form?*).

The pattern is that, as time goes by, science explains more and more and therefore leaves less and less areas of direct responsibility for a deity. Nevertheless, fundamental questions do still remain unanswered. What triggered the Big Bang? What created life on Earth? Currently these have no explanation, offering space for people to believe in an interventionist god. Many scientists see it differently and simply believe that we have not yet found the full answers. It might be thought that, above all, science is about certainty. Building on the success of Kepler and especially of Newton, scientists of the last few centuries certainly formed the opinion that everything could be predicted: if they could measure the position and the movement of every component in a system, then its future behaviour could be calculated precisely. Yet now we know that there is one area of physics in which absolutely certainty is impossible: quantum theory.

Miracles in physics

Quantum theory deals with probable outcomes (see *Are There Alternative Universes?*). Although its equations do not allow impossible things to happen, they do allow highly unlikely things to happen. The theory was developed to describe the behaviour of subatomic particles, the smallest known things in nature. One of the theory's architects, German physicist Werner Heisenberg, discovered in 1926 that at the minuscule scale of subatomic particles, the

Universe is governed by chance not by certainty. Einstein, however, could not accept that chance played any part in the laws of physics. Without a clear law to predict a particle's interaction, Einstein thought that there was too much leeway, too much space that could be filled by an interventionist deity. His antipathy to Heisenberg's idea led to his famous outburst: 'God does not play dice!'

The specific idea in Heisenberg's work that had offended Einstein was that only on the large scale is space a smooth fabric of mass and energy. In the subatomic realm, it is a bubbling mess of particles, which spontaneously form and then disappear in the blink of an eye. Heisenberg found that certain pairs of physical properties, such as time and energy, and position and momentum, were inextricably linked and that the more accurately you measured one, the less accurately you could measure the other. This limit of accuracy has nothing to do with the precision of the equipment used; it is a fundamental uncertainty that is hardwired into the Universe and is known today as 'Heisenberg's Uncertainty Principle'.

To illustrate the link between the position and momentum of a particle, think of a billiard ball rolling across a table in a pitch-black room. To measure its progress, you could roll other balls into the path of the ball and wait for a ricochet. When you hear the balls clip each other, you know that their paths have crossed and so you can say where the original billiard ball is. But in making this detection you have now altered its momentum by knocking it off course and changing its speed. So, although you now know its location, you no longer know its momentum, i.e. where it is going next.

The other pair of interlocked quantities is time and energy. Heisenberg's leap was to realize that, whilst energy must be conserved over a certain period of time, over smaller intervals it can be spontaneously created. This has an astounding consequence because it means that pairs of particles, one of matter and another of antimatter, can suddenly leap into existence from nowhere and then quickly annihilate with their partner again. The uncertainty principle sets the time limit for which they can exist. According to the mathematics, the more energy required to make the particle, that is the more mass it contains, the shorter the period of time for which it can exist. As mind-blowing as it may seem, the uncertainty principle does not just apply to particles. Theoretically, anything can be created in such a 'quantum fluctuation' – rocks, tables, houses, clouds – nothing in the laws of physics forbids this extraordinary behaviour. Even a fully conscious entity could leap into being for a minuscule period of time and, because of the way its brain happened to be 'wired' in that moment, it could even have the illusion of memories and cognition of the Universe.

As crazy as all of this sounds, astronomers and physicists have good reason for believing in the seemingly far-fetched uncertainty principle because, without it, our Sun would not be shining. The temperature of the solar core, even though estimated to be about 16 million degrees, is too low to force hydrogen nuclei to fuse through collisions. Only when the uncertainty principle is folded into the calculations is nuclear fusion possible: because the positions of the hydrogen nuclei are somewhat uncertain, they can be located anywhere within a small radius of the calculated position and so can approach one another closely enough to fuse.

Quantum fluctuations, it seems, are a means of allowing miracles to take place in physics. Other aspects of quantum theory allow particles to behave in even more unexpected ways; certain experimental results can only be understood if the particle involved has somehow been in two places at once. Such discoveries were difficult for scientists to come to terms with; many had been busy trying to explain that the Universe did not need a God at all, that everything could be understood in rational terms. Any highly irrational behaviour seemed to strike against that, perhaps even allow a role for God after all. But, for all its weirdness, the quantum uncertainty did not mean that anything could happen. In an infinite Universe something will inevitably happen even if it has only the tiniest of possibility; but – and this is an important 'but' – if the chances of something happening are zero, that is it is impossible, then not even an infinite length of time will allow it to happen. So the deduction was that the laws of physics must still be used to distinguish what is possible, however unlikely, from the impossible.

God, on the other hand, should be capable of making anything happen, even if the laws of physics forbid it. Although Einstein never liked the idea of quantum uncertainty, he eventually accepted that it did not necessarily provide a niche for God. In later life he hardened his position still further and rejected the idea of God completely. He wrote in 1954, 'The word God is for me nothing more than the expression and product of human weaknesses.'

Fine-tuning

There has been a strong resurgence in the debate amongst cosmologists about whether there is a role for a divine creator in the Universe. It has sprung from indications that the Universe appears to have been designed specifically for human life. As theoreticians have developed a greater understanding of the laws of physics, so they have been able to investigate what the Universe might have been like had the constants of nature been a little different (see *Are There Alternative Universes?*). To their surprise, they have found that the vast majority of possible Universes are not hospitable to life. Some scientists call this conclusion the 'fine-tuning problem' and believe that it needs an explanation.

As an example of fine-tuning, take the rate at which the Universe is expanding. If the expansion rate had been higher than it is today, then matter would have been spread too thinly and the galaxies could not have assembled themselves. At the other end of the scale, if the expansion rate had been too low then the Universe would have collapsed back on itself before stars, planets and humans had time to develop. It is only in a narrow range of values around our current rate of expansion that a Universe with galaxies and stars and planets can form. In general, fine-tuning refers to the apparently slim range of values for the physical constants that can give rise to life, and appears to be telling us that the vast majority of possible Universes are sterile, certainly to life as we recognize it. This leads scientists to wonder why a Universe with human life exists at all, given that it is so highly improbable.

The carbon bottleneck

Perhaps the best example of fine-tuning in the Universe is the so-called 'carbon bottleneck'. Carbon is the element that makes the life-giving DNA molecule possible. Like most of the chemical elements, it is built up from simpler atoms in fusion processes inside stars.

In the 1950s, astronomers used everything they knew about atomic nuclei to develop a scenario for the synthesis of carbon, and called it the 'triple alpha process'. They envisaged three helium nuclei colliding in sequence to build a single nucleus of carbon. However, the final step of the process was proving difficult to understand. British astronomer Fred Hoyle showed that the reaction rates between beryllium and helium, the two final nuclei to react, were spectacularly mismatched. According to the accepted understanding of nuclear reactions, it should be a very unlikely reaction – so unlikely that carbon should be a rare element in space, and yet everywhere astronomers looked they saw relatively large quantities of it. Unless some way could be found to overcome this carbon bottleneck problem, the formation of the carbon needed for life was inexplicable.

Physics was at an impasse. Hoyle, however, was not – he was bold enough to state the obvious. We see the carbon in the Universe, he reasoned, so it must be produced somehow. What's more, carbon is an essential part of life on Earth and has resulted in humans with brains capable of deliberating the problem of the element's formation. So, he concluded, the fact that humans are alive and thinking proved that there was something about the carbon atom that we did not yet understand. He began pondering how nature could speed up the reaction rate between the beryllium and the helium.

Atomic nuclei fuse easily if the two component nuclei have a similar energy state (defined by the internal configuration of the nucleus) to the final product. But, for the known energy states of beryllium and helium, there was no known matching energy state of carbon. Hoyle decided that the fact that he was alive meant that carbon must be capable of holding this quantity of energy, and that all of the experimenters had missed detecting it. The carbon nucleus, he reasoned, would not be able to maintain this energy state for long, otherwise the experimenters would have found it already. So he predicted that although carbon could attain this energy, it must shrug off the excess energy quickly and collapse into the stable form that we see all around us. It need only hold onto the extra energy long enough to make the beryllium–helium reaction possible.

Needless to say, other physicists were highly sceptical, yet when they ran the experiments again, within ten days they had found Hoyle's predicted energy state for carbon. This was the beginning of so-called anthropic (from the Greek word for man) reasoning, which states that the fact that we are alive should inform our reasoning about what is and is not possible in the Universe. In other words, when searching for the laws of physics we must take into account the fact that these laws must ultimately lead to human beings.

Further investigations in this vein have confirmed that the cosmic abundance of elements is a finely tuned network.

Changing the strength of the strong nuclear force, which holds the atomic nucleus together, by just 0.4 percent would destroy this delicate balance and render stars poor factories for carbon. Knowing that life and the Universe are poised on such a knife's edge makes us continue to wonder: why should the formation of the elements be so finely balanced as to make life on Earth possible?

The hand of God?

The short answer to the question of our improbable existence is that no one knows. Some believe the question to be nonsensical because if the Universe had not been like it is, we would not be here to ask the question. Others believe that there must be a profound reason: God made the Universe this way by designing the laws of physics specifically to allow human beings to exist. This would certainly conform to Einstein's non-interventionist God but has an uneasy resonance with the mindset of early 19th century naturalists.

By the early 1800s, human investigation had revealed the most amazing fit between the various life forms and their environments. It was thought to be clear proof that God had designed the world to be perfectly suited for the life it contained. But in 1859 Darwin turned this thinking on its head by presenting his observations that life forms can change with each successive generation to adapt to their surroundings. This idea of natural selection led to the theory of evolution and the belief that the planet and its environment are largely accidental and that life forms evolve by random trial and error – mostly error – to fit whatever niches are available. According to Darwin, the ability to evolve is hardwired into organisms because of an error-prone copying mechanism in the cellular machinery. Whilst some conservative Christian groups continue to believe in 'Intelligent Design', the consensus in the scientific community is that what was once thought of as evidence of the perfection of God is actually imperfect engineering on the molecular level.

So, could the same be true of the Universe at large – that the fine-tuning we see around us is the result of some form of cosmic evolution? This is where the multiverse comes in; if the M-theory landscape is correct (see *Are There Alternative Universes?*), then every possible universe with every possible combination of physical constants is tried out, because there is an infinite number of universes. Inevitably there will be at least one where human life is possible, however finely tuned we need the laws of physics to be. With this view, there is nothing special about our Universe: we just happen to have evolved in the one best suited to the development of our life form; and so there is no cosmological need for God.

The surprising Universe

There is a caveat to keep in mind when discussing fine-tuning and its possible implications – do we perceive that our Universe is fine-tuned for the development of life simply because we lack the imagination to envisage other possibilities? The Universe constantly takes us by surprise, presenting us with wonders that we have neither the wit nor the experience to anticipate. An excellent case in point is the detection of planets around other stars. Astronomers had assumed that the distribution of planets would follow the familiar one of our Solar System: rocky planets close to the Sun and gas giants further out. One team was even collecting data but not analysing it because they were certain that they would need a decade of observations in order to see Jupiter-like planets in their long orbits. In fact, the first detected planets were indeed Jupiter-sized, but orbiting closer to their star than any planet orbits the Sun. The discovery took astronomers completely by surprise, revealing something that they had thought impossible. Could it be that other, previously unthought-of, or discounted, routes to life would be available with other constants of nature? Until we can define what life is (see *Are We Made From Stardust?*), and thus have a concrete rule for what is and is not alive, the discussion of whether we live in a finely tuned Universe is perhaps premature.

Much that we can see and study remains unexplained. The closer we look at the Universe, certainly the more we understand; yet, at the same time, the more mysteries and the more wonder we uncover. Perhaps this will be the pattern of physics forever and there will never be a theory of everything, just an infinite succession of finer and finer details. Or perhaps the final all-embracing theory is just around the corner. Either way, we can be fascinated by cosmology for the journey of discovery that it epitomizes, as well as for the ultimate answers to the big questions that it may deliver.

Glossary

Absolute zero The lowest temperature attainable is -273°C (-459°F), known as zero kelvin (0 K). Essentially, this is the temperature at which atoms cease to exchange energy.

asteroid A small rocky body, 1000 km (600 miles) or less in diameter, orbiting the Sun. Most are found in the asteroid belt between Mars and Jupiter but some have more elliptical orbits and swing past the inner planets.

atmosphere A shell of gases around a planet or moon, of varying combinations and densities.

Big Bang The event that marks the origin of the Universe 13.7 billion years ago: the entire Universe exploded into existence and has been expanding ever since.

billion A thousand million: 10^9 or 1,000,000,000.

black hole An extremely high concentration of mass, with an overwhelming gravitational field from which not even light can escape. Massive stars form black holes when they collapse at the end of their lives, and collapsing gas clouds form supermassive black holes at the centres of galaxies.

comet A chunk of dust and ice, of the order of kilometres wide, left over from the formation of a planetary system. Comets orbiting the Sun are classified according to their orbital period: short-period comets complete their orbit in less than 200 years, while long-period comets can take millions of years.

constants of nature The physical constants that help govern the behaviour of nature, for example the value of the speed of light in a vacuum. Science cannot yet say why they have the values they do.

cosmic microwave background radiation A faint glow of microwave radiation across the entire sky, believed to be the afterglow of the Big Bang itself, now cooled to -270°C (-454°F) or just 2.7 degrees above absolute zero.

cosmological distance ladder A network of techniques that allow astronomers to measure and then estimate distances into the further reaches of the Universe.

dark ages The time between 300,000 years and approximately 1 billion years after the Big Bang when the first celestial objects were forming; before these objects formed, there was no light in the Universe.

dark energy A theoretical energy that permeates all of space, or an unidentified force of nature, employed to explain the acceleration of the expansion of the Universe.

dark matter A hypothetical form of matter thought to outweigh normal atoms by ten times. The effect of its gravity on visible matter infers its presence, but it cannot be detected by emitted radiation.

electron A lightweight, negatively charged particle that usually 'orbits' an atomic nucleus.

fission The act of splitting an atomic nucleus to create two or more lighter atomic nuclei.

fusion The act of joining two atomic nuclei in order to create a heavier nucleus.

galaxy A gravitationally bound system of dust, gas and stars. Galaxies range in size from a few hundred to hundreds of thousands of light years across, and are classified according to their appearance. 'The Galaxy' refers to our own Milky Way.

General Theory of Relativity Albert Einstein's theory that explains gravity as a distortion of space–time.

gravity A physical force that appears to exert a mutual attraction between all masses, governed by the mass and distance apart of the objects.

inflation A postulated moment of intense expansion of space-time shortly after the Big Bang.

kelvin A scale of temperature that places zero as the coldest possible temperature (see absolute zero). To convert from kelvin to Celsius, subtract 273.

late bombardment The final phase of the Solar System's formation, when asteroids and comets pummelled the planets and moons. Our Moon's craters were formed at this time.

light year The distance travelled by light in a vacuum during one year: 9.5 trillion km (5.9 trillion miles); it is used as a measure of astronomical distance to help to keep the vast numbers more manageable.

magnitude A measure of the brightness of a celestial object.

moon A moon is a natural satellite of a planet. There are more than 160 in the Solar System including 'the Moon' which orbits Earth.

multiverse The hypothesised multitude of universes somehow separate from our own.

neutron A heavy, electrically neutral particle that can only exist inside an atomic nucleus.

nucleosynthesis The process by which the chemical elements have been built up into their modern proportions.

orbit The trajectory of one celestial body around another.

parallax A trigonometrical method of measuring the distance of a celestial object.

planet A large object that orbits a star. By historical accident, the name is reserved for the larger bodies, but there is controversy over where the definition of a planet should end. Pluto has recently been re-classified as a dwarf planet.

principle of equivalence The foundation stone of general relativity; it states that acceleration is indistinguishable from a gravitational field.

proton A heavy, positively charged sub-atomic particle that is usually found inside an atomic nucleus.

quantum theory A set of physical laws that describes the behaviour of atoms and sub-atomic particles.

redshift The lengthening of the wavelength of light from a distant object, caused by the expansion of the Universe.

Solar System The Sun and all the bodies in orbit around the Sun: planets and their moons, dwarf planets, asteroids and comets.

space–time The hypothetical fabric of the Universe that, according to the General Theory of Relativity, is deformed by matter and produces gravity.

Special Theory of Relativity Albert Einstein's theory that allows observations from differently moving (but not accelerating) objects to be compared with one another.

star A massive ball of hydrogen and helium bound together by gravity and shining for most of its life with light generated by nuclear fusion.

string theory A possible way of unifying gravity with the other forces of nature, thus providing a single 'theory of everything'. If successful it will supply a quantum theory of gravity.

Sun The star at the centre of our Solar System.

supernova The total destruction of a star by a catastrophic explosion.

trillion A thousand billion: 10^{12} or 1,000,000,000,000.

Universe Everything that exists around us in space. There are estimated to be around 500 billion galaxies in our own Universe and there may be other universes somehow separate from our own.

wormhole A hypothetical shortcut through the fabric of space–time.

Index

Page numbers in **bold** denote an Illustration